T0348831

PHYSICAL TECHNIQUES IN THE STUDY OF
ART, ARCHAEOLOGY AND
CULTURAL HERITAGE

VOLUME 2

Cover illustration: The images printed on the cover are taken from the chapter by Maria Kubik (Chapter 5) and show a plain photograph and two hyperspectral images of a painting in the collection of the Australian War Memorial.

Left: Photograph of the original painting. *Centre and Right:* Hyperspectral images of the painting, showing the location of pigments of different types used by the artist.

[The image of the painting of an Australian soldier by Ivor Hele was taken with the permission of the Australian War Memorial. The photograph of the original painting was taken by the author in the course of her investigations, and is not an official reproduction of the painting (ART40317) in the Australian War Memorial catalogue.]

PHYSICAL TECHNIQUES IN THE STUDY OF
ART, ARCHAEOLOGY AND CULTURAL HERITAGE

Editors

DUDLEY CREAGH
University of Canberra
Faculty of Information Sciences and Engineering
Canberra, ACT 2600, Australia

DAVID BRADLEY
University of Surrey
Department of Physics, Guildford
GU2 7XH, UK

VOLUME 2

ELSEVIER

Amsterdam • Boston • Heidelberg • London • New York • Oxford
Paris • San Diego • San Francisco • Singapore • Sydney • Tokyo

ELSEVIER
Radarweg 29, PO Box 211, 1000 AE Amsterdam, The Netherlands
Linacre House, Jordan Hill, Oxford OX2 8DP, UK

First edition: 2007

ISBN-13: 978-0-444-52856-8
ISSN: 1871-1731

For information on all Elsevier publications
visit our website at books.elsevier.com

Printed and bound by CPI Group (UK) Ltd, Croydon, CR0 4YY

Transferred to Digital Print 2011

**Working together to grow
libraries in developing countries**

www.elsevier.com | www.bookaid.org | www.sabre.org

ELSEVIER   BOOK AID
             International   Sabre Foundation

Contents

Preface

In this Volume 2 of the series on the use of physical techniques for the study of art, archaeology, and cultural heritage, we continue our policy of choosing topics from widely different fields of cultural heritage conservation. Also, we have chosen authors both in their early and late careers.

In Chapter 1, Dudley Creagh writes on "Synchrotron radiation and its use in art, archaeometry, and cultural heritage studies". He is Professor and a Director of the Cultural Heritage Research Centre at the University of Canberra, Canberra, Australia. He has extensive experience in all aspects of cultural heritage research. *Inter alia*, he was a member of the team responsible for the restoration of the Japanese Zero fighter at the Australian War Memorial, conducted research on prestigious medals such as the Victoria Cross and the Lusitania Medal, investigated the effect of self-organizing alkyl chain molecules for the protection of outdoor bronze sculptures, and studied the properties of lubricating oils necessary for the proper preservation of working vintage motor vehicles. Research groups led by him have studied the mechanisms underlying the degradation of Australian aboriginal bark paintings, and examined of the degradation of iron-gall inks on parchment, dyes and pigments in motion picture film, and dyes and pigments on painted surfaces.

Prof. Creagh has also designed new equipment and devised new techniques of analysis. He designed the Australian National Beamline at the Photon Factory, KEK, Tsukuba, Japan. With Dr. Stephen Wilkins, he also designed the unique X-ray diffractometer (BIGDIFF) mounted on it. He designed a number of its accessories, including an eight-position specimen-spinning stage. For surface studies on *air–liquid* interfaces, he designed an X-ray interferometer for the Research School of Chemistry at the Australian National University. He has designed X-ray interferometers that are now finding application in the phase contrast imaging of small objects. More recently, he has designed the infrared beamline for the Australian Synchrotron, Melbourne, Australia. He is currently President of the International Radiation Physics Society.

In continuation of the theme on synchrotron radiation, Loic Bertrand has elaborated, in Chapter 2, on synchrotron imaging for archaeology and art history, conservation, and palaeontology. Dr. Bertrand is the archaeology and cultural heritage officer at the new French synchrotron, Synchrotron Soleil (Orme les Mesuriers, Gif-sur-Yvette, France). He is charged with the task of raising the awareness of cultural heritage scientists to the use of synchrotron radiation for their research. With Dr. Manolis Pantos, he is responsible for the database that lists all the cultural heritage and archaeological publications involving the use of synchrotron radiation. He is an early-career researcher; but mentioning this undervalues the contribution he has already made to the field, using a variety of experimental techniques. In Chapter 2 he describes a number of his activities as well as the research of others.

In the other chapters of this volume, widely different issues are addressed. Chapter 3 is authored by Ivan Cole and his associates Dr. David Paterson and Deborah Lau. This chapter is concerned with the holistic modelling of gas and aerosol deposition, and the degradation of cultural objects. Dr. Cole is the Deputy Chief of the Novel Materials and Processes Division of the Commonwealth Scientific and Industrial Research Organization (Melbourne, Australia). He has over 20 years experience of being involved in projects concerned with the preservation of cultural heritage. Ivan is an internationally recognized leader in the field of life cycle of materials and the development of protective coatings for metals. In 2004, he was a co-winner of the Guy Bengough Award (UK Institute of Materials, Minerals and Mining). He has taken lead roles in major projects in intelligent vehicle health monitoring for aerospace applications, the relation between building design and climate and component life, as well as the development of performance-based guidance standards and codes for durable buildings. He has made a significant contribution in the application of building and material science to the conservation of cultural buildings and collections. Ivan is a member of international and national committees for research and standards in durable structures.

In Chapter 4, Giovanna Di Pietro describes two different types of experiments she has undertaken in the study of the mechanisms underlying the degradation of photographic media. In the first, she describes the degradation of old black-and-white plates. In the second, she outlines her attempts to understand the mechanisms by which the comparatively modern motion picture film degrades. A significant part of this project involved trying to ascertain exactly which dyes were used by Kodak in their motion picture film from about 1980 onwards. The level of secrecy to which this information was protected was great. And, to this day, no information has officially been divulged by the company, although sufficient information has now been acquired to infer the formulations. Giovanna is a post doctoral researcher at the Institute for the Conservation of Monuments, Research Laboratory on Technology and Conservation Polytechnic University of Zurich, Switzerland. Her current project involves monitoring wall paintings using techniques derived from information technology. Giovanna's other research interests include, *inter alia*, the effect of microclimate on canvas paintings. She is a consultant to museums and archives in the field of photographic preservation.

An entirely new technique for the remote investigation of the pigments in paintings is presented by Maria Kubik in Chapter 5. This technique will significantly enhance the ability of conservators to study the palette of pigments used by artists, check for repairs by others, and detect fraudulent paintings. It complements the techniques described by Prof. Franz Mairinger in an earlier Elsevier book *Radiation in Art and Archaeometry*, edited by Creagh and Bradley (2000). Maria is to receive her PhD from the Australian National University in April 2007. She studied conservation in the Cultural Heritage Conservation Course at the University of Canberra, graduating with the degree of Master of Science, specializing in painting conservation. She is at present the Conservator of Paintings at the Western Australia Gallery.

Dudley Creagh
David Bradley

Chapter 1

Synchrotron Radiation and its Use in Art, Archaeometry, and Cultural Heritage Studies

Dudley Creagh

Director, Cultural Heritage Research, Division of Health Science and Design, University of Canberra,
Canberra ACT 2601, Australia
Email: dcreagh@bigpond.net.au

Abstract

Synchrotron radiation has become an increasingly important tool for research in the fields of art, archaeometry, and the conservation of objects of cultural heritage significance. Scientists using conventional laboratory techniques are finding that the fundamental characteristics of synchrotron radiation – high brightness, low divergence, and highly linear polarization – can be used to give information not readily available in the laboratory context. In the author's experience, experiments do not translate directly from the laboratory to the synchrotron radiation laboratory: there are subtle differences in the use of what seem to be similar experimental apparatus. To achieve the best results, the research scientist must be able to discuss his or her research aims meaningfully with beamline scientists. And to be able to do this, the research scientist must have an understanding of the properties of synchrotron radiation, and also the various techniques that are available at synchrotrons but are unavailable in the laboratory. The chapter includes a discussion of synchrotron radiation and its properties, monochromators, detectors, and techniques such as infrared (IR) microscopy; soft X-ray spectroscopy; X-ray diffraction; micro-X-ray diffraction and X-ray fluorescence analysis; X-ray absorption spectroscopy (XAS), including extended X-ray absorption fine structure (EXAFS) and X-ray absorption near edge structure (XANES), and X-ray tomography. The underlying principles of these techniques are discussed here. Later in this book, authors will address these techniques in more detail.

Keywords: Synchrotron radiation, IR microscopy, XRD, micro-XRD, micro-XRF, XAS, XAFS, XANES, X-ray tomography.

Contents

Physical Techniques in the Study of Art, Archaeology and Cultural Heritage
Edited by D. Creagh and D. Bradley

1. INTRODUCTION

Synchrotron radiation is an important scientific tool that is becoming increasingly more useful to scientists working in the fields of archaeology, archaeometry, and the scientific conservation of objects of cultural heritage significance. The growth in the number of peer-reviewed research articles by scientists who are using synchrotron radiation is shown in Fig. 1. The data has been taken from a comprehensive compilation of synchrotron articles that is being made by Drs. Manolis Pantos (SSRC, Daresbury) and Loic Bertrand (Synchrotron-Soleil) can be accessed at http://srsdl.ac.uk/arch/publications.html. Both are

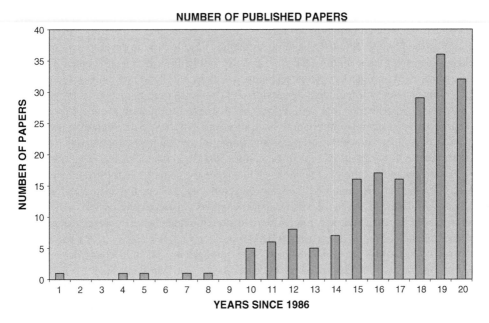

Fig. 1. The growth of peer-reviewed research publications produced by scientists in the fields of archaeology, archaeometry, and cultural heritage conservation since 1986.

contributors to this and later volumes of *Physical Principles in the Study of Art, Archaeology and Cultural Heritage*. The strong growth in publications is mirrored in the increase in the number of workshops held at synchrotron radiation facilities on these topics.

The use of synchrotron by scientists is invariably triggered by the desire to achieve a better understanding of the objects and materials under investigation. And, to a good approximation, the technique chosen is the synchrotron radiation equivalent of a laboratory technique. For example, O'Neill *et al.* (2004) wished to achieve a higher resolution X-ray diffraction pattern from very small amounts of white pigments taken from Australian aboriginal bark paintings than could be achieved using laboratory sources, so that better information could be obtained about the mineral phase composition in the pigments. A laboratory instrument would have required a thousand times more material, and data collection would have taken a hundred times longer. But, as the nature of the problem to be solved became more complex, there arose a need to find other ways to solve the problem – ways that were uniquely suited to the unique properties of synchrotron radiation. The unique properties of synchrotron radiation have enabled the growth of techniques that would not have been feasible in the laboratory situation. A synchrotron radiation source consists of a circulating charged particle beam (usually electrons) in a vacuum vessel (operating vacuum is 10^{-9} mbar) of a high-energy particle accelerator (typically 3 GeV = 3×10^9 eV), and travelling at velocities close to that of light. As will be explained later, radiation is emitted whenever the electron beam is accelerated by the bending magnets that constrain the electron beam to its orbit.

The radiation is highly intense, highly collimated, and highly polarized in the horizontal plane. Also, the emitted radiation covers the whole electromagnetic spectrum: from the far infrared to the hard X region. These unique features have led to the development of many fields of research (XAS, micro-XRD, micro-XRF, and IR microscopy, to name a few) and to the refinement of older laboratory techniques such as XRD and computer-aided tomography. This chapter will include a discussion on synchrotron radiation and its properties.

To devise experiments that will effectively harness the desirable characteristics of synchrotron radiation, it is important to have knowledge of the construction of synchrotron radiation beamlines and of the strengths and limitations of their photon delivery systems. Descriptions will be given of typical beamlines and their monochromators, both of the mirror and single-crystal type, focussing elements, instruments such as diffractometers on which the samples are mounted, and the detectors that collect the scattered radiation. A discussion will be given of such experimental as: infrared microscopy, soft X-ray spectroscopy, X-ray diffraction, micro-X-ray diffraction and X-ray fluorescence analysis, grazing incidence X-ray diffraction (GIXD) and X-ray reflectivity (XRR) techniques, XAS (including XAFS and XANES), and X-ray tomography. The underlying principles of these techniques will be discussed in this chapter. Drs. Bertrand and Pantos will address these techniques in more detail later in this volume, and also in later volumes.

2. THE PRINCIPLES OF SYNCHROTRON RADIATION GENERATION

2.1. Introduction

It is not my intention, in this chapter, to give a full exposition of the principles of synchrotron radiation. That must be reserved for specialized textbooks. See, for example, Atwood (1999), Duke (2000), and Hoffman (2004). Also, Atwood, through the University of California, Berkeley, offers a web-based course on synchrotron radiation (http://www.coe.edu/AST/sxreu).

In this chapter, I shall attempt to present the essence of the subject with little recourse to mathematics. It is assumed that the reader is conversant with the basic notions of electromagnetism. The electromagnetic spectrum arising from the generation of synchrotron radiation ranges from the far infrared (less than 0.1 μm; ~0.1 eV) to hard X-rays (more than 0.1 nm; ~10 keV). The range of interaction is from interactions with atomic and molecular vibrations (far infrared) to crystal diffraction and atomic inner-shell fluorescence effects (X-rays).

The relation between frequency (f), wavelength (λ), and the velocity of light (c) is given by $f\lambda = c$, which can be rewritten as $(h\upsilon)\,\lambda = hc = 1239.842$ eV nm. This expresses the relation in terms of the photon wave packet energy $h\upsilon$. Two useful relations that may assist in understanding some of the figures to follow later are:

- for the energy contained in a photon beam: 1 J = $5.034 \times 10^{15}\,\lambda$ photons (here, λ is the wavelength in nm); and
- for the power in a photon beam: 1 W = $5.034 \times 10^{15}\,\lambda$ photons/s (here, λ is the wavelength in nm).

It is convenient to compare the characteristics of common light sources with those of synchrotron radiation, although at this stage I have not discussed why synchrotron radiation has the properties it has. Table 1 sets out the characteristics of a pearl incandescent bulb, a fluorescent tube used as a replacement for the household incandescent bulb, a typical laboratory laser, and a typical third-generation synchrotron radiation source. It can be seen that the synchrotron radiation source consumes much more source power that the other photon sources.

The photon spectra emitted by both the light bulbs are continuous spectra (although the spectrum of the fluorescent bulb contains the line emission spectrum of the gas used in the bulb). The laser emission is monochromatic, and usually has a small wavelength spread in the emitted line. The synchrotron radiation spectrum is continuous, but, in contrast to the light bulbs that emit in the visible region of the spectrum (less than a decade in wavelength range),

Table 1. Comparison of the characteristics of common light sources (pearl incandescent, bayonet socket fluorescent, common laboratory lasers) with synchrotron radiation sources. The data given is approximate and is given for illustrative purposes only

Characteristic	Incandescent	Fluorescent	Laser	Synchrotron radiation
Source power (W)	100	10	1	10^7
Spectrum	Continuous (0–400 nm)	Continuous (to 400 nm) + discrete spectrum of the fill gas	Monochromatic determined by laser type (1000–400 nm)	Continuous (10 000–0.1 nm)
Source size	Large (2.5×10^3 mm^2)	Large (2.5×10^3 mm^2)	Small (1 mm^2)	Very small (8×10^{-2} mm^2)
Directionality	Omnidirectional	Omnidirectional	Highly directional	Highly directional
Coherence	Incoherent	Incoherent	Coherent	Partially coherent
Polarization	Unpolarized	Unpolarized	Unpolarized	Linearly polarized in horizontal plane mixed polarization off the horizontal plane
Time structure	Continuous	Continuous	Continuous or pulsed	Pulsed

the useful wavelength range of emission is five decades (from far infrared radiation to hard X-ray radiation).

Directionality of emission is an important characteristic of photon sources. Omnidirectional sources emit into all 4π steradians of solid angle. However, in experiments, the experimentalist is usually concerned with illuminating a particular part of their experiment. Let us consider that we wish to illuminate an object 1 mm in diameter, placed 10 m from the source of illumination. Without the addition of optical elements such as focussing mirrors, the fraction of the emission intensity of an omnidirectional source passing through the aperture would be 10^{-8} of the total emission. In contrast, provided the laser was aimed at the aperture, close to 100% of the emitted radiation would pass through the aperture. For a synchrotron radiation source, 100% of the source intensity would pass through the aperture.

Source size is important in two respects. The smaller the source size, the brighter the source is said to be. Also, the size source has an effect on the intensity of the beam at a distance from the source. It is convenient here to introduce definitions related to photon transport that will be used throughout this chapter. They are:

- Flux (F): the number of photons passing a unit area per unit time;
- Brightness (B): photon flux per unit source area per unit solid angle.

Nothing has been said here about the wavelength of photon radiation. For continuous radiation, a slice of the spectrum is taken, usually 0.1% of the bandwidth. When referring to a particular radiation, the definitions of flux and brightness are modified to be:

- Spectral flux (photons/mm^2/s/0.1% BW),
- Spectral brightness (photons/mm^2/mrad^2/s/0.1% BW),
- Spectral flux per unit solid angle (photons/mrad^2/s/0.1% BW), and
- Spectral flux per unit horizontal angle (photons/mm^2/mrad/s/0.1% BW).

Note that, according to Liouville's theorem, flux and brightness are invariant with respect to the propagation of photons through free space and linear optical elements. They are the best descriptors of source strength.

Coherence is related to the ability of radiation emitted from different parts of the source to have fixed passes in relation with one another. Longitudinal coherence length is defined as

$$L_{coh} = \frac{\lambda^2}{2\Delta\lambda}$$

Thus, an optical laser has high coherence since λ is typically 633 nm and $\Delta\lambda$ is less than 0.1 nm ($L_{coh} = 2 \times 10^6$ nm), whereas a light bulb would have typically 600 nm and $\Delta\lambda$ is perhaps 800 nm ($L_{coh} = 225$ nm). The question of coherence in the case of synchrotron radiation is not quite so straightforward: it depends on the method of production of the synchrotron radiation.

Of the radiation sources, only synchrotron radiation sources produce polarized radiation. Synchrotron radiation is normally 100% linearly polarized in the plane of the electron orbit, but, as the view of the radiation changes, so does the degree of linear polarization.

The current in the synchrotron has a time structure arising from the fact that the electrons are injected into the storage ring in bunches spaced from one another by a long time period, compared to the length of the bunch. Typically, an electron bunch may be 50 ps long, and the spacing between bunches may be 2.4 ns. This fact can be used in studying fast atomic and molecular reactions.

Also, the beam intensity at any beamline will decrease as a function of time after injection. Collisions with residual gas atoms and molecules in a high-vacuum system, which takes some of the electrons away from the electron trajectory, and radiative losses at the bending magnets, which cause a change in position of the electron beam, reduce the current in the ring. This means that the radiated power decreases after injection occurs. Thus, the intensity of the beam may fall, say, 20%, in the course of a day. Therefore, more electrons have to be injected to return the intensity to its original value at the end of a day. Or, alternatively, electrons are added to the storage ring at regular time intervals to maintain the storage ring current at its nominated value. Whatever the strategy taken, for accurate measurements, the incident beam intensity must be monitored.

2.2. Synchrotron radiation sources

2.2.1. Bending-magnet sources

The earliest dedicated synchrotron radiation sources (referred to as first-generation sources) consisted of a circular vacuum chamber with a central radius (R) into which the electrons were injected at an energy (E_e), and a large electromagnet that provided a uniform magnetic induction (**B**) chosen such that the injected electron beam returned to complete the circular trajectory. First-generation synchrotrons were either small "tabletop" systems with radii less than 10 m and operating in the infrared or soft X-ray regions, or were the result of radiation from very high-energy nuclear physics installations such as DESY (Hamburg, Germany), where hard X-ray beams were available, but the characteristics and availability of the beams were determined by the requirements of the nuclear structure program (the so-called parasitic mode of operation). Figure 2 shows schematically the arrangement of a tabletop synchrotron. Some of this class of synchrotrons are still in operation. The Helios synchrotron manufactured by Oxford Instruments was installed initially in Singapore in around 1995, and had a superconducting magnet to produce the magnetic induction. The magnetic induction for this first-generation synchrotron radiation source is perpendicular to the plane of the electron orbit and must be constant over the whole area of the vacuum chamber. Under these conditions, each of the electrons in the beam experiences a constant force (**F**), which produces acceleration (a) towards the centre of the circular orbit.

$$\mathbf{F} = -ev \times \mathbf{B} = ma$$

and

$$R = \frac{mv}{e\mathbf{B}}$$

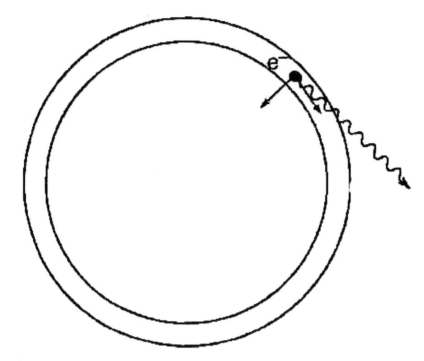

Fig. 2. Schematic representation of a first-generation synchrotron radiation source, showing the circular electron orbit, photon radiation, and acceleration of the electrons. The magnetic induction is constant and directed into the plane of the paper.

Because an accelerated charge radiates electromagnetic radiation, each electron produces a dipole radiation field (Fig. 3(a)). Photons are emitted uniformly over the horizontal plane, which includes the orbit. In this case, the electron is assumed to be travelling at a lesser velocity than the velocity of light.

The electrons, however, are injected at a velocity close to that of light with the consequence that the dipole field carried in the electron's frame becomes distorted when viewed by an observer because of a time compression, given by

$$\frac{dt}{dt'} = 1 - \beta \cos \theta$$

where θ is the deviation of the angle of viewing relative to the tangent to the electron orbit, and $\beta = (1/2\gamma^2) - 1$.

Here, $\gamma = m_e/m_0$ where m_e is the mass of the electron and m_0 is its rest mass (≈ 0.511 MeV). If, for example, the electron energy is 500 MeV, $\gamma \approx 1000$, and a significant relativistic Doppler shift exists, as is shown schematically in Fig. 3(b). Note that this Doppler shift is strongly angle dependent. The intensity of the radiation distribution is constrained to within a cone of angle approximately $(1/\gamma)$, as illustrated in Fig. 3(c). The ratio γ is used as a measure of the directionality of the synchrotron radiation beam.

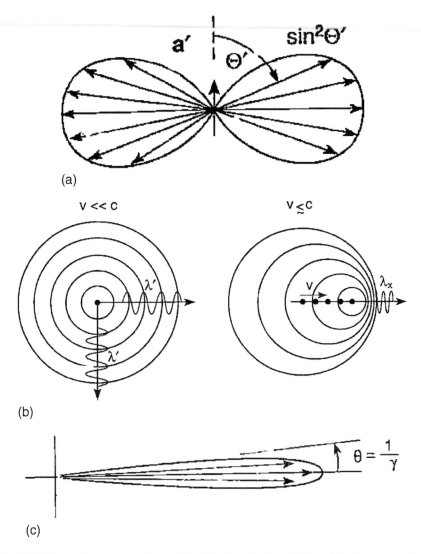

Fig. 3. (a) Schematic representation of the dipole radiation field emitted by an accelerated electron. The electron is assumed to be travelling at a lower velocity than that of light. (b) Schematic representation of the relativistic Doppler shift due to relativistic time compression. (c) Schematic representation of the dipole radiation field emitted by a relativistic accelerated electron. The field is restricted to a cone of angle $(1/2\gamma)$.

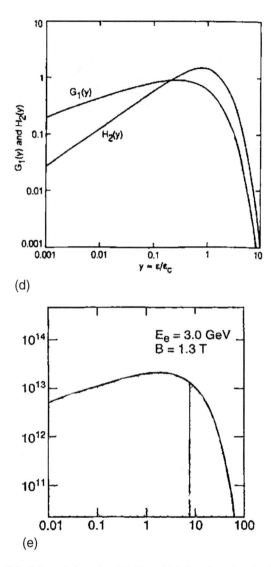

(d)

(e)

Fig. 3. (d) The modified Bessel function $H_2(y)$, which is related to the flux per solid angle, and the function $G_1(y)$, which is related to the flux per horizontal angle (the flux in the vertical plane has been integrated) as a function of y (= E/E_c = ω/ω_c). (e) The spectral distribution emanating from a bending magnet at the Australian Synchrotron. Here, B = 1.3 T for the dipole magnet, the synchrotron energy E_c = 3 GeV, $\Delta\omega/\omega$ = 0.1%, and $\Delta\theta$ = 1 mrad.

The emission of radiation covers a wide energy range. The critical energy (E_c) of the synchrotron radiation source is defined as the median energy of the energy range: *i.e.* half the possible energies lies below E_c, and half above.

$$E_c = \left(\frac{h}{2\pi}\right)\omega_c = \frac{3eh\mathbf{B}\gamma^2}{4\pi m}$$

where e is the electron charge, m is its mass, and ω_c is the critical frequency. The expressions that relate to the production of the synchrotron radiation spectrum are complicated, and contain modified Bessel functions, the form of which determines the shape of the spectral distribution.

The photon flux per unit solid angle (θ is the horizontal angle measured from the tangent to the viewing point and ψ is the horizontal angle measured from the tangent to the viewing point) is, in the horizontal plane ($\psi = 0$)

$$\frac{d^2 F}{d\theta d\psi} = \left(\frac{3\alpha}{4\pi^2}\right)\gamma^2\left(\frac{\Delta\omega}{\omega}\right)\left(\frac{I}{e}\right)H_2\left(\frac{\omega}{\omega_c}\right)$$

where I is the circulating current and

$$H_2\left(\frac{\omega}{\omega_c}\right) = \left(\frac{\omega}{\omega_c}\right)K_{2/3}^2\left(\frac{\omega}{2\omega_c}\right)$$

In this formula, I have grouped similar quantities: all the constants are groups, the angular frequencies are grouped, and the circulating current and electronic charge are grouped.

$K_{2/3}^2(\omega/2\omega_c)$ is a modified Bessel function of the second kind; α is the fine structure constant ($\approx 1/137$). The modified Bessel function is shown graphically in Fig. 3(d). Also plotted on this graph is the function $G_1(\omega/\omega_c)$, which is the integral of $H_2(\omega/\omega_c)$ over the vertical angle ψ. These curves are "universal curves" for synchrotron radiation emanating from bending magnets. Figure 3(e) shows the spectral distribution emanating from a bending magnet at the Australian Synchrotron. Here, $B = 1.3$ T for the dipole magnet, the synchrotron energy $E_c = 3$ GeV, $\Delta\omega/\omega = 0.1\%$ and $\Delta\theta = 1$ mrad.

Thus far, I have not discussed the details of how electron beams are generated, accelerated to high energy, and stored. Figure 4(a) is a plan of the Australian Synchrotron. It shows, commencing from the inner part of the plan, that electron bunches are generated in a linear accelerator, in which they are accelerated to an energy of 100 MeV. Figure 4(b) shows a technician installing the RF driver coils around the LINAC tube. The electron bunches are then directed into a booster synchrotron in which the current in the dipole magnets is increased, and RF energy is applied so as to increase the energy of the electron bunch to 3 GeV. Figure 4(c) shows a section of the booster synchrotron in its shielding tunnel. The box-like objects are dipole magnets surrounding the vacuum vessel in which the electron bunches circulate. Note that synchrotron radiation is generated at each of these

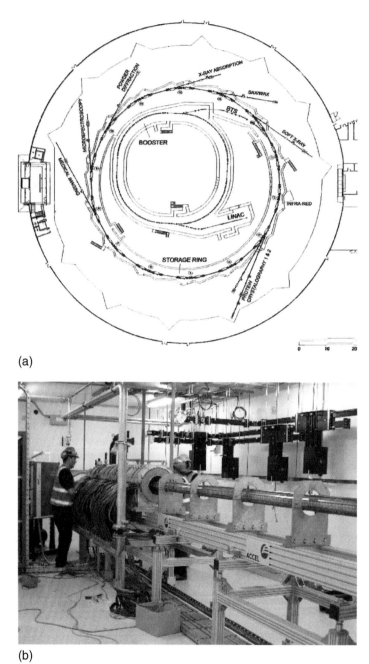

(a)

(b)

Fig. 4. (a) Floor plan of the Australian Synchrotron. The LINAC accelerates electrons from a thermionic source to 100 MeV. These electrons then pass into a booster synchrotron that accelerates the electron bunches to 3 GeV. These are then diverted into the storage ring where the electron bunches circulate, producing radiation whenever the bunches are accelerated. (b) Assembling the LINAC at the Australian Synchrotron. Electron bunches are generated thermionically and accelerated by the linear accelerator to 100 MeV.

(c)

(d)

Fig. 4. (c) Dipole magnets surrounding the vacuum chambers in the booster synchrotron for the Australian Synchrotron. The electrons are accelerated by an applied radiofrequency field. As they gain energy, the field strength in the dipole magnets is increased to maintain the electrons in their orbit. (d) A view of a dipole magnet before it is rolled into position over the vacuum chamber (the curved section on the right of the picture).

(e)

Fig. 4. (e) A closer view of the dipole magnet and its associated sextupole magnet (on the left of the dipole magnet), and the quadrupole magnet (on the right of the dipole magnet), which are integral components of the storage ring lattice.

dipole magnets, and is the major source of energy loss from the accelerator. When the electron bunch energy has reached 3 GeV, a "kicker magnet" diverts the electrons in the booster synchrotron into the synchrotron storage ring. This ring is not circular, but consists of straight sections connected to curved sections at which the bending magnets are situated. Additional magnets, a magnetic quadrupole and a magnetic sextupole, are mounted at the entrance and exit ports of each dipole magnet to steer and focus the electron bunches. Figure 4(d) shows the positioning of a dipole magnet in the storage ring. Two dipole magnets exist between neighbouring straight sections; in this case, there are 14 pairs of dipole magnets. At the right of the picture, the vacuum chamber can be seen. The dipole magnet is about to be rolled into position around the vacuum chamber. A closer view of the dipole magnet and its associated sextupole magnet is shown in Fig. 4(e). A magnetic quadrupole is situated at the far end of the dipole magnet.

2.2.2. Second- and third-generation synchrotrons

Second-generation synchrotrons such as the Photon Factory, Tsukuba, Japan, were built for dedicated use by scientists, and consist of straight vacuum sections, at the ends of which are placed bending magnets. In the straight sections, insertion devices, *i.e.* devices in which the electron beam is perturbed from the normal orbit, can be placed. These insertion devices comprise a set of magnets of alternating polarity that have the effect of decreasing the radius of curvature of the electron beam, thereby changing the emission characteristics of the source. These periodic magnetic arrays are referred to as undulators if the radiation emitted by successive bends adds coherently, or as wigglers if the radiation emitted by successive bends adds incoherently. Wigglers, undulators, and their properties will be discussed in detail later.

Third-generation synchrotrons are the adaptation of second-generation synchrotrons to produce small electron beam size and divergence, and the straight sections are optimized for the inclusion of insertion devices. Figure 5(a) shows schematically the organization of a modern synchrotron radiation facility, showing how the electron beam may be modified to produce radiation of different characteristics. Third-generation synchrotrons have their optics arranged so as to produce electron beams of small size, and the dimensions of the electron beam source are referred to as σ_x and σ_y for the beam sizes in the horizontal and the vertical directions, respectively. The amount of radiation collected in the horizontal plane by an experiment depends on the size of the exit apertures in the horizontal plane. The effective emission angle in the vertical plane is limited to $\pm 1/\gamma$ of the horizontal plane, and is approximately

$$\sigma_\psi = \left(\frac{570}{\gamma}\right)\left(\frac{E}{E_c}\right)^{0.43}$$

The machine characteristics of the Australian Synchrotron are given in Table 2. The emittance is a measure of the intrinsic source size of the synchrotron radiation storage ring.

In insertion devices, the electrons travel through a periodic linear magnetic structure. In such a structure, the magnetic induction may be devised to be sinusoidal and be oriented normal to the plane of the electron orbit, such that

$$B(z) = B_o \cos\left(\frac{2\pi z}{\lambda_u}\right)$$

where λ_u is the wavelength of the magnetic array. This imposes a sinusoidal motion on the electron, and this is constrained to the horizontal plane. This is illustrated schematically in Fig. 5(b). An important parameter describing the motion of the electron is the deflection parameter K ($= eB_o\lambda_u/2\pi mc = 0.934 \lambda_u B_o$). In terms of K, the maximum angular deflection from the orbit is $\delta = K/\gamma$.

For $K \leq 1$, radiation from the bends can interfere with one another because the excursion of the electrons lies within the $1/\gamma$ limit for the radiation cone. This particular structure gives rise to undulator radiation.

For $K \gg 1$, interference effects are not of importance, and the radiation that emanates from this structure is referred to as wiggler radiation.

2.2.2.1. Wiggler radiation. K is usually a large number (>10) for periodic magnetic arrays designed to emit wiggler radiation. In this case, the radiation from different parts of the electron trajectory add incoherently, and the total flux from the array is $2N$ times the appropriate formula for a bending magnet, with the values of B and R taken at the point of the

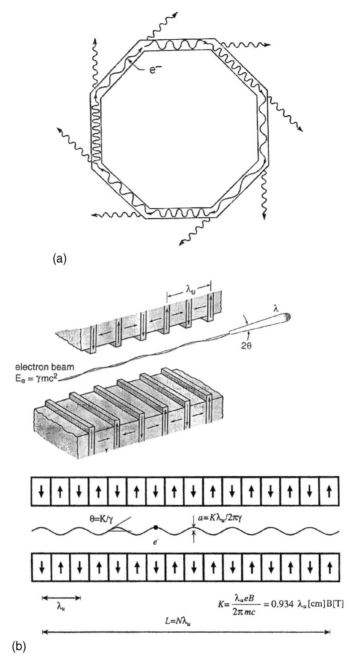

(a)

(b)

Fig. 5. (a) Schematic representation of the organization of a modern synchrotron radiation facility, showing how the electron beam may be modified to produce radiation of different characteristics by the use of insertion devices in the straight sections to modify the trajectory of the electron bunches and the use of bending magnets to divert them in their path. (b) Electron motion within a periodic magnetic field. The schematic diagram is for an undulator. These are linear periodic magnetic arrays (situated either inside or outside the storage ring vacuum vessel) in which the radiation from the bends add together constructively to produce a coherent radiation pattern.

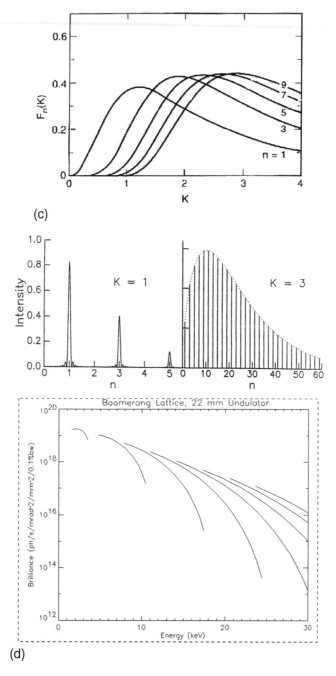

Fig. 5. (c) Plot of $F_n(K)$ as a function of K. (d) Relative intensity plots showing the effects of K on a harmonic number. Shown, as well, are the characteristics of an undulator chosen for use in the spectromicroscopy beamline at the Australian Synchrotron.

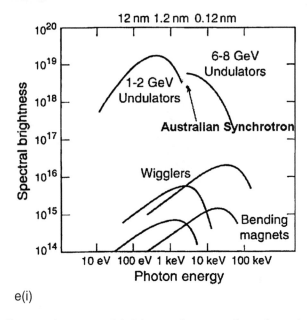

e(i)

Fig. 5. (e) (i) Comparative spectral brightness for a number of synchrotron radiation sources. At the bottom are spectral brightness curves for bending-magnet sources with two different values of E_c, one a soft X-ray source and the other a hard X-ray source. Above them are plots of the spectral brightness for wigglers mounted in straight sections of the two storage rings. Note that the radiation from the wigglers is similar in shape to the bending-magnet sources, but $2N$ times more intense, and shifted to higher energies. At the top are curves relating to the spectral brightness of undulators in the straight sections of a low-energy and a high-energy ring. The dot indicates the peak intensity of radiation from and undulator at the Australian Synchrotron.

Table 2. Machine and electron beam parameters for the Australian Synchrotron (0.1 m dispersion optics)

Parameter	Value
Energy	3.0 GeV
Circumference	216 m
Current	200 mA
Horizontal size	181 μm (1 sigma)
Vertical size (1% coupling)	45 μm (1 sigma)
Horizontal divergence	197 μrad
Vertical divergence	23 μrad
Bending-magnet field	1.3 T
Emittance	7.33 nm rad

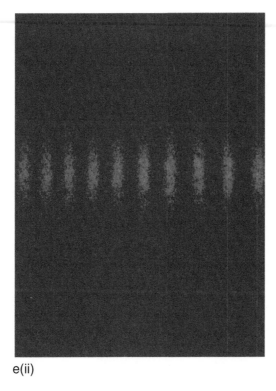

e(ii)

Fig. 5. (e) (ii) Streak photograph showing the time structure of the synchrotron radiation beam.

electron trajectory, tangential to the direction of observation. Here, N is the number of magnetic periods. For a horizontal angle θ,

$$E_c(\theta) = E_{cmax}\left(1 - \left(\frac{\theta}{\delta}\right)^2\right)^{0.5}$$

where $E_{cmax} = 0.665\ E^2 B_o$ (E_o is expressed in GeV and B_o in Tesla).

In the horizontal plane ($\psi = 0$), the radiation is linearly polarized in the horizontal direction. As in the case of simple bending-magnet radiation, the direction of polarization changes as the ψ changes, but because the elliptical polarization from one half period combines with the elliptical polarization with the next half period (which has the opposite sense), the resultant polarization remains linear.

2.2.2.2. Undulator radiation. Figure 5(b) is a schematic representation of an undulator insertion device. Note that a diagram for a wiggler would look the same: the periodic magnetic array would look similar, but the deflection parameter K ($= eB_o\ \lambda_u/2\pi mc = 0.934\ \lambda_u B_o$) is

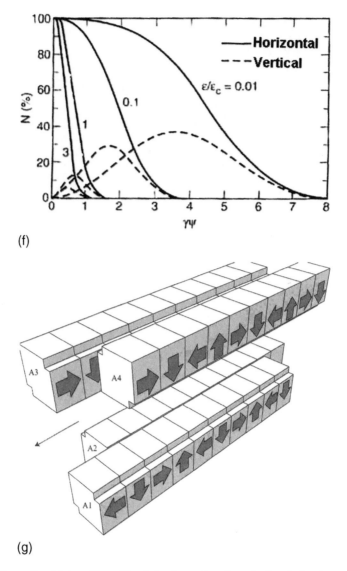

(f)

(g)

Fig. 5. (f) Normalized intensities of both horizontal and vertical polarization components, calculated as functions of the product of the vertical angle of observation ψ and $\gamma\,(=E/m_ec^2)$. E is the energy of the electron. E_c is the critical energy, defined as $0.665\ E^2B$, where E is expressed in GeV and B is given in Tesla. (g) Schematic representation of a helical undulator (after Chavanne, 2002). There are four blocks of magnets in the arrays (A1, A2, A3, and A4) for every undulator period. The directions of the magnetization within the structures are shown. By moving two opposing magnet arrays with respect to the other two, the field strengths of the components of the vertical and horizontal magnetic fields can be varied. This changes the phase relations between the two impressed oscillations, thereby changing the polarization of the electron beam.

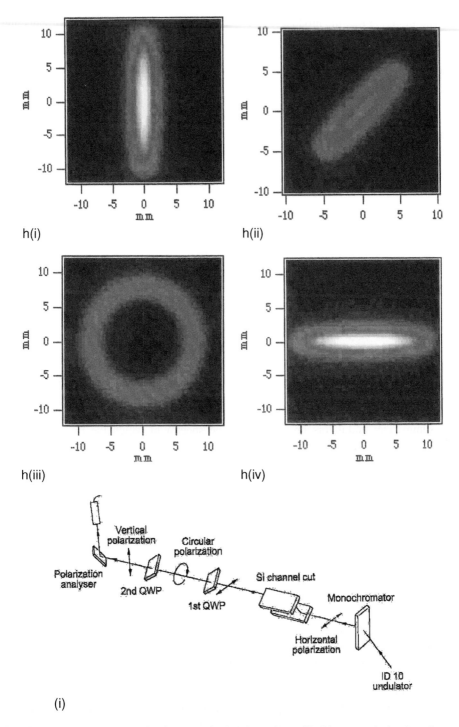

Fig. 5. (h) (i) Linear polarization, vertical orientation. (ii) Linear polarization from a helical undulator inclined at 45° to vertical. (iii) Circular polarization.

different, implying that the product $\lambda_u B_o$ is different. K is greater than 1 for wigglers and less than 1 for undulators. The wavelength of the fundamental on the axis is

$$\lambda_1 = \left(\frac{\left(1 + \dfrac{K^2}{2} \right)}{2\gamma^2} \right) \lambda_u$$

and the relative bandwidth of the nth harmonic is

$$\left(\frac{\Delta\lambda}{\lambda} \right) \approx \left(\frac{1}{nN} \right)$$

The on-axis peak intensity of the nth harmonic is zero for n = even numbers and for n = odd numbers

$$\left(\frac{\mathrm{d}^2 F_n}{\mathrm{d}\theta \mathrm{d}\psi} \right) = 1.744 \times 10^{14} N^2 E^2 I F_n(K)$$

where E is expressed in GeV and I in Amperes. $F_n(K)$ is the sum of Bessel functions and is plotted in Fig. 5(c) for the first nine harmonics. The relative effect on the spectral distribution for two different values of K ($K = 1$; $K = 3$) is shown in Fig. 5(d).

To summarize: undulators provide quasi-line spectra in which the lines have high brightness, relatively small breadth, and small angular divergence in the forward direction.

Fig. 5, cont'd (h) (iv) Linear polarization, horizontal orientation. (i) Schematic diagram of a beamline designed to deliver circularly polarized light to a sample. Linearly polarized radiation from an undulator passes through two monochromators, the first a Laue-type (transmission) monochromator, and then a double-crystal Bragg (reflection) monochromator. The radiation remains linearly polarized in the horizontal plane. It then passes through a quarter wave plate (λ/4), which is oriented so as to produce equal amounts of vertical and horizontal polarization in the wavefields within the quarter-wave plate (QWP). On leaving the QWP, these wavefields combine to give circularly polarized radiation. This can then be used to irradiate a sample, for example, a layer of self-organized alkyl chains in a lubricating oil on a metal surface, to determine the orientation of the alkyl chains. Analysis of the resultant scattered radiation can be effected using another QWP to determine the amount, say, of vertical or horizontal polarization exists in the beam.

The position of the lines can be controlled by varying the gap between the poles of the magnet.

2.2.2.3. Comparison of spectral brightness of synchrotron radiation sources. Figure 5(e) shows the comparative spectral brightness of the three different synchrotron radiation sources. At the bottom are spectral brightness curves for bending-magnet sources with two different values of E_c, one a soft X-ray source and the other a hard X-ray source. Above them are plots of the spectral brightness for wigglers mounted in straight sections of the two storage rings. Note that the radiation from the wigglers are similar in shape to the bending-magnet sources, but $2N$ times more intense, and shifted to higher energies. At the top are curves relating to the spectral brightness of undulators in the straight sections of a low-energy and a high-energy ring. The dot indicates the peak intensity of radiation from and undulator at the Australian Synchrotron.

2.2.2.4. Polarization of synchrotron radiation beams. The polarization of synchrotron radiation beams can be manipulated either at the source, or by the insertion of phase plates in the beamline.

Polarization created at the source. Synchrotron radiation sources give rise to radiation that is linearly polarized in the plane of the orbit for bending-magnet, wiggler, and undulator sources. This arises from the fact that the magnetic field directions in all of these cases in perpendicular to the plane of orbit. If, however, the source is viewed off the axis, changes of polarization are observed (Fig. 5(f)). This shows the normalized intensities of both horizontal and vertical polarization components, calculated as functions of the product of the vertical angle of observation ψ and $\gamma (=E/m_ec^2)$. E is the energy of the electron. E_c is the critical energy, defined as $0.665\ E^2B$, where E is expressed in GeV and B is given in Tesla. For $E/E_c = \gamma = 1$, for example, the polarization changes from 100% linearly polarization horizontal radiation on the axis ($\gamma\psi = 0$) to 0% at around $\psi = 1.6$. The degree of vertically polarized radiation rises from 0% on axis to reach a maximum of around 17% at $\psi = 0.6$. Thus, for a particular photon energy, it is possible to see different admixtures of polarizations by changing the viewing angle. The intensity of the mixed polarized light is significantly lower than that of the on-axis linearly polarized light.

The ability to produce photon beams with a particular polarization state has been acquired with the invention of a new class of undulators. This class of undulators is referred to as Apple II undulators (Sasaki, 1994; Chavanne, 2002). An Apple II undulator has recently been installed at the Daresbury Laboratory (Hannon *et al.*, 2004), and one has recently been commissioned for the Australian Synchrotron.

Apple II undulators are referred to as helical undulators because the electron beam traverses two orthogonal periodic magnetic fields, usually constructed from permanent magnets such as NdFeB or Sm_2Co_{17}. There are four arrays of magnets (A1, A2, A3, and A4) (Fig. 5(g)). There are four blocks of magnets in the arrays (A1, A2, A3, and A4) for every undulator period. The directions of magnetization within the structures are shown. By moving two opposing magnet arrays with respect to the other two, the field strengths of the components of the vertical and horizontal magnetic fields can be varied. This changes the phase relations between the two impressed oscillations, thereby changing the polarization of the electron beam. Figure 5(h) (i–iv) shows various polarization states that can be attained by varying the position of the moveable magnetic array. In order, they are, respectively: linear polarization, vertical orientation; linear polarization inclined 45° to vertical; circular polarization; and linear polarization, horizontal orientation.

Polarization created by optical elements.   Phase shifts can be induced in a monochromatic X-ray beam by using reflections from single-crystal silicon crystals used in a Laue (transmission) configuration. In simple terms, the incoming radiation stimulates coupled wavefields having both parallel and perpendicular components in the crystal. The degree of polarization is determined by the thickness, orientation, and reflection type (111, 333, 311, and so on) (see Giles *et al.*, 1994). In Fig. 5(i) a schematic diagram of a beamline designed to produce circularly polarized light is shown. Linearly polarized radiation from an undulator passes through two monochromators: the first, a Laue-type (transmission) monochromator, and then a double-crystal Bragg (reflection) monochromator. The radiation remains linearly polarized in the horizontal plane. It then passes through a quarter wave plate ($\lambda/4$), which is oriented so as to produce equal amounts of vertical and horizontal polarization in the wavefields within the quarter wave plate (QWP). On leaving the QWP, these wavefields combine to give circularly polarized radiation. This can then be used to irradiate a sample, for example, a layer of self-organized alkyl chains in lubricating oil on a metal surface, to determine the orientation of the alkyl chains. Analysis of the resultant scattered radiation can be effected using another QWP to determine the amount, say, of vertical or horizontal polarization that exists in the beam.

3. SYNCHROTRON RADIATION BEAMLINES

3.1. General comments

In Section 2, the characteristics of synchrotron radiation were described. In what follows, the various elements that may comprise the photon delivery system for a particular experimental apparatus are described. In general, all beamlines have to be held under high vacuum ($\approx 1 \times 10^{-7}$ mbar), and fast gate valves are provided to isolate the experiment and the beamline from the ultrahigh vacuum of the electron storage ring. Infrared, vacuum ultraviolet, and soft X-ray beamlines operate under the same vacuum conditions as the storage ring ($\approx 1 \times 10^{-9}$ mbar). For these beamlines, the usual technique for isolating the high vacuum from the atmosphere, beryllium windows, cannot be employed, because the beryllium is opaque to the radiation required for the experiments.

There are many configurations of beamlines: they are usually tailored to meet the particular needs of experimental scientists. In this section, I shall describe some of the more common configurations that are of general use for conservation scientists: those for infrared microscopy, microspectroscopy, XRR, X-ray diffraction, XAS, XAFS and XANES, and X-ray imaging. To commence with, however, a generic beamline that incorporates the elements used to produce the beam to be used in an experiment will be described. This is an X-ray beamline, and, as shown schematically in Fig. 6(a), capable of delivering a focussed beam to the sample. The details of the pipes, maintained at high vacuum, through which the X-ray beam passes, the vacuum isolating valves, and the experimental hutch that

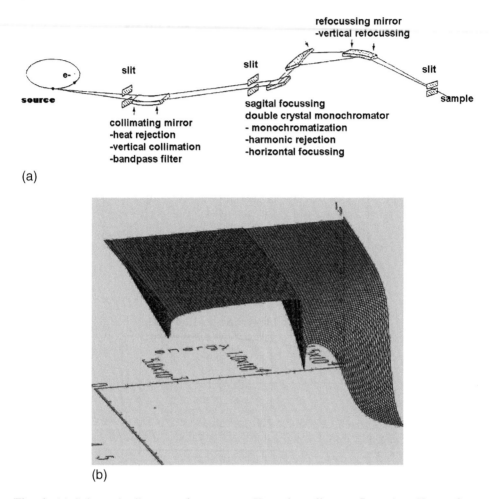

(a)

(b)

Fig. 6. (a) Schematic diagram of a common X-ray beamline configuration. X-rays from the synchrotron radiation source pass through a beam-defining slit and impinge on a mirror. In practice, this mirror may be flat or concave upwards. The latter is chosen if a parallel beam is required. The beam then passes through a slit placed to minimize unwanted scattered radiation from the mirror and falls on a double-crystal silicon mono-chromator. Using Bragg reflection, a particular photon energy (wavelength) can be selected by the first crystal from the broad spectrum reflected by the focussing mirror. The Bragg reflected beam, however, may contain harmonics. These are eliminated by slightly detuning the second crystal. If focussing of the beam following the monochromator is required, the second crystal can be bent sagittally. To complete the focussing and produce a spot on the specimen, a refocussing mirror is used. This can also redirect the beam to a certain extent. (b) The reflectivity of copper as a function of energy (0–15 keV) and angle (0.1–5°). Note that the reflectivity at 0.1° is close to 1, but there are irregularities at two energies that correspond to the absorption edges for copper ($K\alpha = 8.797$ keV; $L_{group} = 1.096, 0.952, 0.932$ keV). For a given energy, as the glancing incidence is increased, the reflectivity falls, and the effect of the absorption edge increases. For high energies, pene-tration (transmission) of the beam into the copper surface dominates over reflection.

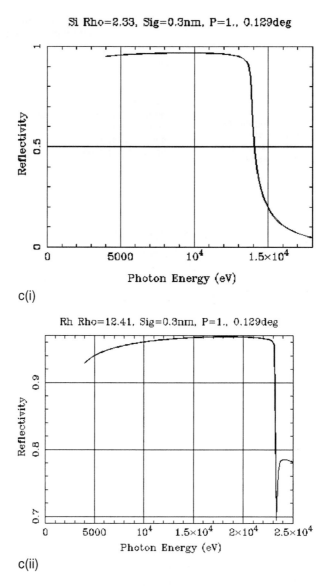

c(i)

c(ii)

Fig. 6. (c) (i) Reflectivity of silicon as a function of photon energy for an incident angle of 0.129° and surface roughness 0.3 nm. (ii) Reflectivity of rhodium as a function of photon energy for an incident angle of 0.129° and surface roughness 0.3 nm.

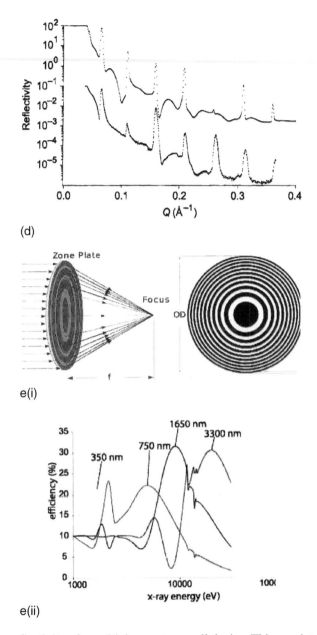

(d)

e(i)

e(ii)

Fig. 6. (d) The reflectivity of a multiple quantum well device. This consists of 40 alternating layers of AlGaAs and InGaAs, each 10 nm thick. The bottom curve is a curve calculated using the electromagnetic theory to show clearly the extent to which it agrees with the experiment. Multilayer quasi-Bragg reflection devices can be fabricated by alternating very thin layers of a heavy element (tungsten) with a light element (silicon). These devices obey the Bragg equation ($2d \sin \theta = n\lambda$) where d is the spacing of the tungsten layers. (e) (i) Schematic representation of the operation of a hard X-ray zone plate. (e) (ii) The energy performance of a hard X-ray zone plate. Note that efficiency is a strong (and irregular) function of X-ray energy. For X-ray energies greater that 10 keV, a zone plate 3300 nm long will have about 20% efficiency.

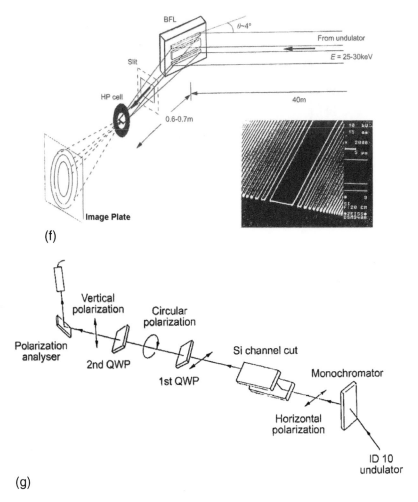

(f)

(g)

Fig. 6. (f) The Bragg–Fresnel Zoneplate system used to produce vertical focussing to complement Bragg reflection of radiation from an undulator. This produces a small focal spot at the high pressure cell. The diffraction pattern is observed using imaging plates. (g) A system for manipulating the polarization of the linearly polarized undulator beam to produce circularly polarized light using quarter-wave plates.

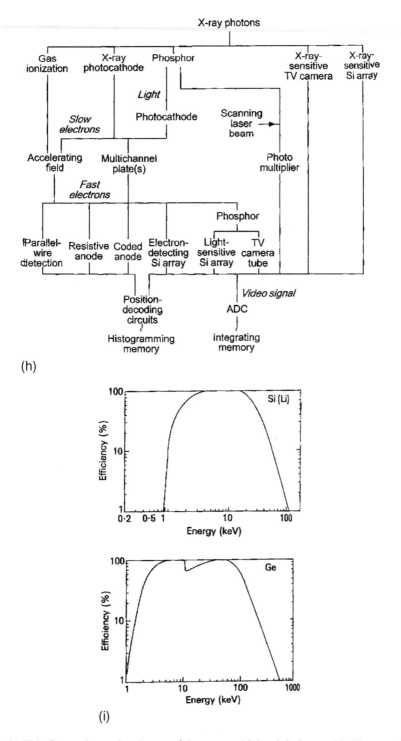

Fig. 6. (h) This figure shows the classes of detectors which might be used in X-ray analytical equipment. (i) Energy response curves for SiLi and HPGE detectors.

houses the experimental apparatus are not shown. For X-rays, the radiation level in the proximity of the specimen precludes the presence of an operator. All alignment, control, and experimental procedures must be effected remotely by the experimenter.

In Fig. 6(a), X-rays from the synchrotron radiation source pass through a beam-defining slit, and impinge on a mirror. In practice, this mirror may be flat or concave upwards. As I shall describe later, the X-rays are totally externally reflected by the mirror, and the incident beam makes an angle of <1° with the surface. If a parallel beam in the vertical plane is required, the mirror is bent to be concave upwards. The beam then passes through a slit places to minimize unwanted scattered radiation and falls on a double-crystal silicon monochromator. Using Bragg reflection, a particular photon energy (wavelength) can be selected by the first crystal from the broad spectrum reflected by the focussing mirror. The Bragg reflected beam, however, may contain harmonics. These are eliminated by slightly detuning the second crystal. If focussing of the beam following the monochromator is required, the second crystal can be bent sagitally. To complete the focussing to produce a spot on the specimen, a refocussing mirror is used. This can also redirect the beam to an extent determined by the angle of total external reflection.

In what follows, I shall describe in some detail the properties of mirrors, monochromators, and other focussing elements used in synchrotron radiation beamlines. In the course of this, I shall introduce concepts that will later be used in the discussion of X-ray experiments.

3.1.1. Interfaces

All sources of X-rays, whether produced by conventional sealed tubes, rotating anode system, or synchrotron radiation sources, emit over a broad spectral range. In many cases, this spectral diversity is of concern, and techniques have been developed to minimize the problem. As mentioned earlier, these involve the use of filters, mirrors, and Laue and Bragg crystal monochromators, chosen so as to provide the best compromise between flux and a spectral purity in a particular experiment. This section does not purport to be a comprehensive exposition on the topic of filters and monochromators. Rather, it seeks to point the reader towards the information given elsewhere.

The ability to select photon energies, or bands of energies, depends on the scattering power of the atoms from which the monochromator is made, and the arrangement of the atoms within the monochromator. In brief, the scattering power of an atom (f) is defined, for a given incident photon energy, as the ratio of the scattering power of the atom to that of a free Thomson electron. See, for example, Creagh (2004 a,b) and Sections 4.2.4 and 4.2.6. The scattering power is denoted by the symbol $f(\omega, \Delta)$ and is a complex quantity, the real part of which, $f'(\omega, \Delta)$, is related to elastic scattering cross section, and the imaginary part of which, $f''(\omega, \Delta)$, is related directly to the photoelectric scattering cross section and, therefore, the linear attenuation coefficient μ_l. The attenuation of X-rays is given by

$$I = I_\mathrm{o} \exp(-\mu_l t)$$

where t is the thickness of the material through which the X-rays travel.

At an interface between, say, air and the material from which the monochromator is made, reflection and refraction of the incident photons can occur, as dictated by Maxwell's equations. There is an associated refractive index n given by

$$n = (1 + \chi)^{0.5}$$

where

$$\chi = -\frac{r_e \lambda^2}{\pi} \Sigma_j \, N_j \, f_j(\omega, \Delta)$$

and r_e is the classical radius of the electron and N_j is the number density of atoms of type j.

An angle of total external reflection α_c exists for the material, and this is a function of the incident photon energy since $f_j(\omega, \Delta)$ is a function of photon energy. In Fig. 6(b), calculations of the reflectivity of copper as a function of energy (0–15 keV) and angle (0.1–5°) are shown. Note that the reflectivity at 0.1° is close to 1, but there are irregularities in two regions that correspond to absorption edges for copper ($K\alpha$ = 8.797 keV; $L_{1,2,3}$ = 1.096, 0.952, 0.932 keV). For a given energy, as the glancing incidence is increased, the reflectivity falls, and the effect of the absorption edge increases. For high energies, penetration into copper dominates over reflection. Thus, a polychromatic beam incident at the critical angle of one of the photon energies (E) will reflect totally those components having energies less than E, and transmit those components with energies greater than E.

Figure 6(c) shows calculations for the reflectivity of single layers of silicon and rhodium as a function of incident energy for a fixed angle of incidence (0.129°). As the critical angle is exceeded, the reflectivity varies as E^{-2}. The effect of increasing atomic number can be seen: the higher the atomic factor $f(\omega, \Delta)$, the greater the energy that can be reflected from the surface.

Interfaces can therefore be used to act as low-pass energy filters. The surface roughness and the existence of impurities and contaminants on the interface will, however, influence the characteristics of the reflecting surface, sometimes significantly.

3.1.2. Mirrors and capillaries

3.1.2.1 Mirrors. Although neither of these devices is, strictly speaking, monochromators, they nevertheless form component parts of monochromator systems in the laboratory and in synchrotron radiation sources.

In the laboratory, they are used in conjunction with conventional sealed tubes and rotating anode sources, the emission from which consists of bremsstrahlung, upon which the characteristic spectrum of the anode material is superimposed. The shape of the bremsstrahlung spectrum can be significantly modified by mirrors, and the intensity emitted at harmonics of the characteristic can be significantly reduced. More importantly, the mirrors can be fashioned into shapes that enable the emitted radiation to be brought to a focus. Ellipsoidal, logarithmic spiral, and toroidal mirrors have been manufactured commercially for use in laboratory X-ray sources. Since the X-rays are emitted isotropically from the anode surface, it is important to devise a mirror system that has a maximum angle of acceptance and a relatively long focal length.

At synchrotron radiation sources, the high intensities that are generated over a very broad spectral range give rise to significant heat loading of subsequent monochromators, and therefore degrade the performance of these elements. In many systems, mirrors are used as the first optical element in the monochromator to reduce the heat load on the primary monochromator, and to make it easier for the subsequent monochromators to reject harmonics of the chosen radiation. Shaped mirror geometries are often used to focus the beam in the horizontal plane (see Fig. 6(a)). A schematic diagram of the optical elements of a typical synchrotron radiation beamline is shown in Fig. 6(a). In this beam-line, the primary mirror acts as a thermal shunt for the subsequent monochromator, mini-mizes the high-energy component that may give rise to possible harmonic content in the final beam, and focusses the beam in the horizontal plane. The radius of curvatures of mirrors can be changed using mechanical four-point bending systems (Oshima *et al.*, 1986). More recent advances in mirror technology enable the shape of the mirror to be changed through use of the piezoelectric effect (Sussini and Labergerie, 1995).

3.1.2.2. Capillaries. Capillaries and bundles of capillaries are finding increasing use in situations where a focussed beam is required. The radiation is guided along the capillary by total external reflection, and the shape of the capillary determines the overall flux gain and the uniformity of the focussed spot. Gains in flux of 100 and more have been reported. However, there is a degradation in the angular divergence of the outgoing beam. For single capillaries, applications are laboratory-based protein crystallography, microtomography, X-ray microscopy, and micro-X-ray fluorescence spectroscopy. The design and construc-tion of capillaries for use in the laboratory and at synchrotron radiation sources has been discussed by Bilderback *et al.* (1994), Balaic *et al.* (1995, 1996), and Engstrom *et al.* (1996) They are usually used after other monochromators in these applications, and their role as a low-pass energy filter is not of much significance.

Bundles of capillaries are currently being produced commercially to produce focussed beams (ellipsoid-shaped bundles) or half-ellipsoidal shaped bundles to form beams of large cross section from conventional laboratory sources (Kumakhov, 1990; Kumakhov and Komarov, 1990; Peele *et al.*, 1996; Owens *et al.*, 1996).

3.1.2.3. Quasi-Bragg reflectors. For one interface, the reflectivity and transmissivity of the surface is determined by the Fresnel equations, *i.e.*:

$$R = \left| \left(\frac{\theta_1 - \theta_2}{\theta_1 + \theta_2} \right) \right|^2$$

$$T = \left| \left(\frac{2\theta_1}{\theta_1 + \theta_2} \right) \right|^2$$

where θ_1 and θ_2 are the angles between the incident ray and the surface plane, and the reflected ray and the surface plane, respectively.

If a succession of interfaces exists, the possibility of interference between successively reflected rays exists. Parameters that define the position of the interference maxima, the line breadth of those maxima, and the line intensity depends *inter alia* on the regularity in layer thickness, the interface surface roughness, and the existence of surface tilts between successive interfaces. Algorithms for solving this type of problem are incorporated in software currently available from a number of commercial sources (Bede Scientific, Siemens, and Philips). The reflectivity profile of a system having a periodic layer structure is shown in Fig. 6(d). This is the reflectivity profile for a multiple quantum well structure of alternating aluminium gallium arsenide and indium gallium arsenide layers (Holt *et al.*, 1996). Note the interference maxima that are superimposed on the Fresnel reflectivity curve. From the full-width-half-maximum of these interference lines, it can be inferred that the energy discrimination of the system, $\Delta E/E$, is 2%. The energy range that can be reflected by such a multilayer system depends on the interlayer thickness: the higher the photon energy, the thinner the layer thickness. Multilayer quasi-Bragg reflection devices can be fabricated by alternating very thin layers of a heavy element (tungsten) with a light element (silicon). These devices obey the Bragg equation ($2d \sin \theta = n\lambda$, where d is the spacing of the tungsten layers). They can have high integrated reflectivities compared to monocrystalline materials, but their energy resolution is poor.

Commercially available multilayer mirrors exist, and hitherto have been used as monochromators in the soft X-ray region in X-ray fluorescence spectrometers. These monochromators are typically made of alternating layers of tungsten and carbon to maximize the difference in scattering length density at the interfaces. Although the energy resolution of such systems is not especially good, these monochromators have a good angle of acceptance for the incident beam, and reasonably high photon fluxes can be achieved using conventional laboratory sources.

A recent development of this, the Goebel mirror system, is supplied as an accessory to a commercially available diffractometer (Siemens, Osmic™, 1996). This system combines the focussing capacity of a curved mirror with the energy selectivity of the multilayer system. The spacing between layers in this class of mirror multilayers can be laterally graded to enhance the incident acceptance angle. These multilayers can be fixed to mirrors of any figure to a precision of 0.3 arc minutes, and can therefore can be used to form parallel beams (parabolic optical elements) as well as focussed beams (elliptical optical elements) of high quality.

3.2. Monochromators

3.2.1. Crystal monochromators

Strictly monochromatic radiation is impossible since all atomic energy levels have a finite width, and emission from these levels is therefore spread over a finite energy range. The corresponding radiative linewidth is important for the correct evaluation of the dispersion corrections in the neighbourhood of absorption edge. Even Mossbauer lines, originating as they do from nuclear energy levels that are much narrower than atomic energy levels, have finite linewidths. Achieving linewidths comparable to these requires the use of monochromators that utilize carefully selected single crystal reflections.

Crystal monochromators make use of the periodicity of "perfect" crystals to select the desired photon energy from a range of photon energies. This is described by Bragg's law:

$$2d_{hkl} \sin\theta = n\lambda$$

where d_{hkl} is the spacing between the planes having Miller indices hkl, θ is the angle of incidence, n is the order of a particular reflection ($n = 1, 2, 3...$), and λ is the wavelength.

If there are wavelength components with values near $\lambda/2$, $\lambda/3$, etc., these will be reflected, as well as the wanted radiation and harmonic contamination that can result. This can be a particular difficulty in spectroscopic experiments, particularly XAFS, XANES, and DAFS.

The Bragg equation neglects the effect of the refractive index of the material. This is usually omitted from Bragg's law since it is on order of 10^{-5} in magnitude. Because the refractive index is a strong function of wavelength, the successive harmonics are reflected at angles slightly different from the Bragg angle of the fundamental. This fact can be used in multiple-reflection monochromators to minimize harmonic contamination. Most X-ray beamlines incorporate a double-crystal monochromator, and almost all have the capability of harmonic rejection by the slight defocussing of the second element. Rotations of around $2''$ of arc are usually used to eliminate the third and higher harmonics in silicon double-crystal monochromators.

Each Bragg reflection has a finite linewidth, the Darwin width, arising from the interaction of the radiation with the periodic electron charge distribution (see, for example, Warren (1968)). Each Bragg reflection, therefore, contains a spread of photon energies. The higher the Miller indices, the narrower the Darwin width becomes. Thus, for experiments involving the Mossbauer effect, extreme back reflection geometry is used at the expense of photon flux.

If the beam propagates through the specimen, the geometry is referred to as transmission geometry, or *Laue* geometry. If the beam is reflected from the surface the geometry is referred to as reflection geometry, or *Bragg* geometry. Bragg geometry is most commonly used in the construction of crystal monochromators. Laue geometry has been used in only a relatively few applications, until recently. The need to handle high photon fluxes with their associated high power load has led to the use of diamond crystals in Laue configurations as one of the first components of X-ray optical systems (Freund, 1993). Phase plates can be created using Laue geometry (Giles *et al.*, 1994). A schematic diagram of a system used at the European Synchrotron Radiation Facility is shown in Fig. 6(g). Radiation from an insertion device falls on a Laue geometry pre-monochromator and passes through a channel-cut (multiple-reflection) monochromator. The strongly linear polarization from the source and the monochromators can be changed into circular polarization by the asymmetric Laue geometry polarizer, and analysed by a similar Laue geometry analysing crystal.

3.2.2. Laboratory monochromator systems

Although this chapter deals with synchrotron radiation sources, it would also be useful to discuss laboratory monochromator systems here. In practice, synchrotron radiation beamlines are used to supplement data gained from laboratory sources, and it is necessary to understand the strengths and weaknesses of both systems. Many laboratories use powder diffractometers using the Bragg–Brentano configuration. For these, a sufficient degree of monochromatization is achieved through the use of a diffracted-beam monochromator consisting of a curved graphite monochromator and detector mounted on the 2θ arm of the diffractometer. Such a device rejects the unwanted $K\beta$ radiation and fluorescence from the sample, with little change in the magnitude of the $K\alpha$ and $K\alpha_2$ lines. In other uses, incident beam monochromators are used to produce closely monochromatic beams of the desired energy.

For most applications, this simple means of monochromatization is adequate. Increasingly, however, more versatility and accuracy are being demanded of laboratory diffractometer systems. Increased angular accuracy in the θ and 2θ axes, excellent monochromatization, and parallel beam geometry are all demands of a user community using improved techniques of data collection and data analysis. The necessity to study thin films has generated a need for accurately collimated beams of small cross section, and there is a need to have well-collimated and monochromatic beams for the study of rough surfaces. This, coupled with the need to analyse data using the Rietveld method (Young, 1993), has caused a revolution in the design of commercial diffractometers, with the use of principles long since used in synchrotron radiation research for the design of laboratory instruments.

3.2.3. Multiple-reflection monochromators for use with laboratory and synchrotron radiation sources

Single-reflection devices produce reflected beams with quite wide quasi-Lorentzian tails, a situation that is not acceptable, for example, for the study of small-angle scattering (SAXS). The effect of the tails can be reduced significantly through the use of multiple Bragg reflections.

The use of multiple Bragg reflections from a channel cut in a monolithic silicon crystal such that the channel lies parallel to the (111) planes of the crystal was shown by Bonse and Hart (1965) to remove the tails of reflections almost completely. This class of device, referred to as a *(symmetrical) channel-cut crystal*, is the most frequently used form of monochromator produced for modern X-ray laboratory diffractometers and beamlines at synchrotron radiation sources.

The use of symmetrical and asymmetrical Bragg reflections for the production of highly collimated monochromatic beams has been discussed by Beaumont and Hart (1974). This article contains descriptions of the configurations of channel-cut monochromators, and combinations of channel-cut monochromators used in modern laboratory diffractometers produced by Philips, Siemens, and Bede Scientific. In a later article, Hart (1971) discussed the whole gamut of Bragg reflecting X-ray optical devices. Hart and Rodriguez (1978) extended this to include a class of devices in which the second wafer of the channel-cut monochromator could be tilted with respect to the first, thereby providing an offset of the crystal rocking curves, with the consequent removal of most of the contaminant harmonic radiation The version of monochromator shown here is designed to provide thermal stability for high incident photon fluxes. Berman and Hart (1991) have also devised a class of adaptive X-ray monochromators to be used at high thermal loads where thermal expansion can cause a significant degradation of the rocking curve, and therefore a significant loss of flux and spectral purity. The cooling of Bragg geometry monochromators at high photon fluxes presents a difficult problem in design.

Modern double-crystal monochromators have very complicated first-crystal structures, designed to minimize the effect of thermal stress on the Bragg reflecting planes. Because the lattice coefficient of expansion is close to zero at liquid nitrogen temperatures, the first crystals in many high-brilliance systems are cooled using liquid nitrogen. These systems are to be found at all third-generation storage rings. Systems are available from a number of manufacturers: for example, Oxford Danfysik (http://www.oxford-danfysik.com/), Accel (http://www.accel.de/), and Kohsa Seiki (http://www.kohzueurope.com/pdf/ssm.pdf).

Kikuta and Kohra (1970) and Kikuta (1971) have discussed in some detail the performance of *asymmetrical channel-cut* monochromators. These find application under circumstances in which beam widths need to be condensed or expanded, in X-ray tomography, or for micro-X-ray fluorescence spectroscopy. Hashizume (1983) has described the design of asymmetrical monolithic crystal monochromators for the elimination of harmonics from synchrotron radiation beams.

Many installations use a system designed by the Kohzu Company as their primary monochromator (http://www.kagaku.com/kohzu/english.html). This is a separated element design in which the reference crystal is set on the axis of the monochromator and the first crystal is set so as to satisfy the Bragg condition in both elements. One element can be tilted slightly to reduce harmonic contamination. When the wavelength is changed (*i.e.* θ is changed), the position of the first wafer is changed either by mechanical linkages or by electronic positioning devices so as to maintain the position of the outgoing beam in the same place as it was initially. This design of a fixed-height separated element monochromator was due initially to Matsushita *et al.* (1986). More recent designs incorporate liquid nitrogen cooling of the first crystal for use with high-power insertion devices at synchrotron radiation sources. In many installations, the second crystal can be bent into a cylindrical shape to focus the beam in the horizontal plane. The design of such a sagittally focussing monochromator is discussed by Stephens *et al.* (1992). Creagh and Garrett (1995) have described the properties of a monochromator based on a primary monochromator (Berman and Hart (1991)) and a sagittally focussing second monochromator at the Australian National Beamline at the Photon Factory.

3.3. Focussing optical elements

Fresnel lens zone plates are being used with increasing frequency in fields such as protein crystallography, microspectroscopy, microtomography, and microdiffraction for the focussing of hard X-rays. A number of commercial suppliers are now producing zone plates for use at synchrotron radiation sources. See, for example, http://www.xradia.com. A Fresnel zone plate is constructed from the concentric rings of a highly absorbing material such as gold, supported on a lightly absorbing substrate. Microlithographic techniques are often used in the fabrication. The focal length of a zone plate such as that shown schematically in Fig. 6(e)(i) is given by

$$F = \left(\frac{D\Delta R_n}{\lambda} \right)$$

where D is the diameter of the zone plate and ΔR_n is the width of the outermost zone. The resolution of the zone plate is determined by the Rayleigh criterion ($=1.22\ \Delta R_n$).

Photon throughput is dependent on the thickness of the gold layers and the X-ray energy (Fig. 6(e)(ii)). The efficiency of a 3300-nm-thick zone plate is better than 20% between 10 and 60 keV.

A recent innovation in X-ray optics has been made at the European Synchrotron Radiation Facility by the group led by Snigirev (1996). This combines Bragg reflection of X-rays from a silicon crystal with Fresnel reflection from a linear zone plate structure lithographically etched on its surface. Hanfland *et al.* (1994) have reported the use of this class of reflecting optics for the focussing of 25–30-keV photon beams for high-pressure crystallography experiments (Fig. 6(f)).

As can be seen in Fig. 6(a), focussing can be effected by using curved (concave towards the incident beam) mirrors. Where two such mirrors, vertically focussing and horizontally focussing, are situated in close proximity to one another, they are said to be a Kirkpatrick–Baez pair.

3.4. Polarization

All scattering of X-rays by atoms causes a probable change of polarization in the beam. Jennings (1981) has discussed the effects of monochromators on the polarization state for conventional diffractometers of that era. For accurate Rietveld modelling or for accurate charge density studies, the theoretical scattered intensity must be known. This is not a problem at synchrotron radiation sources, where the incident beam is initially almost completely linearly polarized in the plane of the orbit, and is subsequently made more linearly polarized through Bragg reflection in the monochromator systems. Rather, it is a problem in the laboratory-based systems, where the source is in general a source of elliptical polarization. It is essential to determine the polarization for the particular monochromator and the source combined to determine the correct form of the polarization factor to use in the formulae used to calculate scattered intensity. Please see Section 2.2.2.4.

4. DETECTORS

A comprehensive review of the detectors available for the detection of X-rays has been given by Amemiya *et al.* (2004). As can be seen in Fig. 6(g), the number of types of detectors to choose from is large. In practice, only five different classes of detectors are used: ionization chambers, microstrip proportional detectors, scintillation detectors, solid state detectors, and CCD array detectors.

Table 3 sets out some of the characteristics of these detectors. The details of these detectors are not discussed here. These details can be found by accessing the websites of the manufacturers.

4.1. Ionization chambers

Ionization chambers are essentially gas-filled chambers through which the X-ray beam passes. Two parallel electrodes within the chamber are arranged to be parallel to the X-ray beam. The separation of the plates is fixed, but the fill gas can be changed (usually He, N_2, or A, with CH_4, C_2H_4, or CO_2 as the quench gas), as can the gas pressure (typically 0.7–1.3 atm). The processes that occur when the X-ray beam interacts with the gas can be described macroscopically by Paschen's law: the breakdown characteristics of a gas are a function (generally not linear) of the product of the gas pressure and the gap length, usually written as $V = f(pd)$, where p is the pressure and d is the gap distance. In reality, pressure should be replaced with gas density. The voltage to be used in ionization has to be less than what would initiate a spark. In the microscopic description, the electron ejected from the atom by the X-ray is accelerated by the applied field. Three distinctly different types of behaviour can occur, depending on the applied field.

- If the field is low, the electrons are accelerated towards the anode, and if there has been no significant scattering by gas atoms, the number of electrons striking the anode is equal to the number of X-ray photons interacting with the gas atoms. The current is measured and converted to a frequency that is then counted by a high-frequency counter.
- If the electrons have sufficient energy to ionize the gas atoms with which they interact, further secondary ionization can occur, and a charge cloud is formed. This is referred to as the Townsend secondary ionization effect (the Townsend α coefficient). The multiplication of electrons in this process can be large, and can produce charge pulses, the height of which is proportional to the incident X-ray flux. This is a photon counting system. The size of the pulse for a given fill gas depends on the electric field and the quantity of quench gas present.
- If the charge cloud becomes too large, a steady state discharge or, more likely, a spark will occur. Whether the spark persists to become an arc depends on the capacity of the external circuitry to maintain the field. The quenching of the arc is assisted by the presence of gases such as CO_2 and CH_4. When a spark occurs, the system is said to be operating in the Geiger counter region.

Because the fill gas in ionization chambers interacts weakly with the incident beam, they are used extensively:

- for monitoring the intensity of the incident beam (necessary because the intensity of radiation emitted by a storage ring decreases as a function of time due to the interaction of the circulating electron bunches with residual gas in the vacuum chambers); and
- in transmission X-ray absorption fine structure (XAFS) experiments (Section 5.5).

Table 3. Characteristics of the various types of X-ray detectors in common use at synchrotron radiation facilities. Only typical values are quoted

Characteristics	Ionization	Proportional	Scintillation	Solid State	CCD array	Imaging plate
Detecting medium	Gas	Gas	Solid	Solid	Solid	Solid
Energy range (keV)	4–100[a]	4–20[b]	4–100[c]	1–20 (SiLi)[d] 1–100 (HPGE)	1–100[e]	1–100[f]
Type of detection	Integrating[g]	Photon[h]	Photon	Photon	Photon	Integrating
Energy discrimination ($\Delta E/E$)	None	0.05[i]	0.25	0.025 (SiLi) 0.05 (HPGE)	0.05	None

[a] The energy range depends on the fill gas (He, N2, A), the quench gas (CO_2, CH4), and the operating pressure. A typical ionization chamber might operate with 90% A and 10% CH_4 at atmospheric pressure. The applied voltage is sufficient only to produce a small ionization current flow.

[b] The energy range depends on the fill gas (A, Xe), the quench gas (CO_2, CH_4), and the operating pressure. A typical gas proportional detector might operate with 95% A and 5% CH_4 at several times atmospheric pressure. The applied voltage is sufficient only to produce gas multiplication proportional to the energy deposited in the chamber by the incident photon.

[c] These are two-part devices comprising a scintillation material (plastic, NaI(Tl), ZnS, CsI, $CdWO_4$) in which the incident photon causes ionization in the scintillator, and, when the atom relaxes to its ground state, a photon of energy proportional to the energy of the incident photon is released. This then interacts with a phosphor coating on the second part of the device (either a photomultiplier tube or an avalanche photodiode).

[d] These are two-part devices comprising a block of silicon doped with lithium (SiLi) or high-purity germanium (HPGE) in which the incident photon causes ionization in the semiconductor, and this cloud of ionization is swept to the anode by an applied voltage. The number of electrons in the current is proportional to the energy of the incident photon. This current pulse is converted to a voltage pulse and is then amplified using standard nucleonic techniques. The energy range is determined by the window on the device (usually 1–3-mm thick Be) and the photonic band structure of the material.

[e] CCD array detectors are similar to the scintillation detectors.

[f] Imaging plates are constructed similar to a conventional camera film, in that a storage phosphor is coated onto a polymer backing using an organic binder. This storage phosphor consists of small grains (< 5 μm) of BAFBr:Eu^{2+}.

[g] Integrating detectors accumulate the charge deposited by the incident photons.

[h] Photon detectors detect and amplify individual photon interactions.

[i] This depends on the fill gas, the operating pressure, and the detected photon energy. The standard test energy is the emission line, Fe $K\alpha$ (6.40 keV).

Continued

Table 3. Characteristics of the various types of X-ray detectors in common use at synchrotron radiation facilities. Only typical values are quoted—cont'd

Characteristics	Ionization	Proportional	Scintillation	Solid State	CCD array	Imaging plate
Maximum count rate (Hz)	NA	20 000	1 000 000	20 000	20 000	NA
Dynamic range^j	10^5	10^4	10^4	10^4	10^4	10^5
Spatial resolution (μm)	NA	≈300	NA	NA	100^k	10^l
Detection efficiency (%)^m	Very low	20	80	≈100^n	≈100°	≈100
Detector aperture (mm × mm)	50×10^p	$2 \times L^q$	10 (diameter)	10 (diameter)	0.1×0.1 (pixel size)	400×200^r

^{j} Dynamic range is the range over which the received data is proportional to the received photon flux.

^{k} This depends on the pixel size in the array.

^{l} This depends of the grain size (<5 μm).

^{m} Specified in the operating range.

^{n} HPGE has its absorption edges at around 11 keV. In the region 10–30 keV, detector efficiency is around 85%. This is for an individual germanium detector. For a germanium detector array (10, 30, and 100 element detectors are available), the overall performance depends on the packing of the detector elements.

^{o} The overall detection efficiency of the detector depends on the packing of the pixels, the coupling of the array to its amplifiers, and the readout scan rate of the array.

^{p} The active length of an ionization chamber can be from 10 to 300 mm.

^{q} The length of a strip detector can range from 10 mm for an individual detector to 300 mm for a strip detector. The strip detector is used as the detector to replace the moving 2θ detector on a standard diffractometer.

^{r} This figure is for an individual pixel. Because of the packing of pixels, the overall performance of an array will be less than that quoted for an individual pixel.

4.2. Proportional detectors

Proportional detectors operate in the second regime of the ionization process of a gas. It is essential to arrest as many X-ray photons as possible within the detector. Therefore, they commonly use heavy inert gases such as xenon or argon at high pressures (3–10 atm) to maximize the stopping power of the detector.

Although they are frequently used in laboratory systems as individual detectors, in synchrotrons they are usually used in the form of position-sensitive detectors. Designed for simultaneous data collection, these can reduce data collection time by a factor of 100–1000 because the detector remains stationary in the scattered beam path rather that having to be moved across the beam path. In small-angle scattering experiments, linear proportional detectors are often used to measure the intensity distribution of a region of $10°$ or more about the direct beam position with good spatial resolution. This detector is ideally suited for phase transformation studies as a function of temperature and stress analysis. There is a loss in spatial resolution, and the detector can be saturated by intense X-ray beams, however. Two types of detectors are in common use: linear (wire) and curved (or blade) detectors. Two-dimensional detectors (wire) do exist, but they are not common.

In curved detectors, a blade-shaped curved anode, which has a sharp edge of a few tenths of a micron radius and an intense electric field are applied. To maximize mechanical stability, the anode is designed as a very thin curved blade. The X-ray photons ionize the gas atoms in the detector. The electrons are immediately accelerated, and additionally have sufficient energy to ionize other gas atoms. A very fast multiplication phenomenon appears, the so-called "avalanche". On the cathode readout strips, an inducted charge is produced perpendicular to the impact point of the avalanche. The position of this charge is determined electronically. The charge travels to the left and to the right along this delay line to both ends of the line. The difference in arrival times of the charge at each end correlates to the position of the avalanche on the anode.

Curved-position sensitive detectors are used, for example, as elements integrated into X-ray powder diffraction systems where they are usually curved into an arc to replace the 2θ detector in powder diffractometers. They are good for studying time-resolved diffraction experiments because their effective 2θ range is $120°$. However, their angular resolution is only around $0.03°$; less than that of a laboratory diffractometer. And there are limitations to the maximum count rate that can be observed (see, for example, http://www.italstructures.com/detectors.htm).

4.3. Scintillation detectors

Scintillation detectors are widely used in X-ray beamlines. Essentially, they comprise a scintillation medium to which a photomultiplier tube is coupled. In some instances, the photomultiplier tube is replaced by an avalanche photodiode. The incoming photon is stopped with material in the scintillator crystal, causing ionization. The amount of ionization created is proportional to the energy deposited by the photon. Recombination of electrons in the ionized atoms causes the release of a pulse of light. In the case of the photomultiplier tube, the photon pulse activates a phosphor at the cathode of the photomultiplier tube, and the secondary electrons emitted from this are accelerated through a number of dynodes. Current amplifications of 10^5 or more are possible. At the anode, the current pulse passes into a resistor, and the resulting voltage pulse is processed by the following electronics. This may include some kind of multichannel amplifier (mca), a device for plotting the number of received photons having a particular energy as a function of that energy. Since the voltage received is proportional to the photon energy, the scintillation detector system displays the incident photon spectrum. Avalanche photodiodes behave in a way similar to photomultipliers.

The materials used as scintillators must:

- be transparent,
- stop the incident photon (which implies that the atomic mass of the atoms in the material is large),
- produce short flashes of light (so that count rates can be high),
- have a small linewidth, and
- have a linear response to incident photon energy.

Plastic scintillators usually have poor stopping power and energy resolution. They can, however, receive high incident beam fluxes.

Most scintillators are made from alkali halides (NaI(Tl), CsI) or tungstates (CdWO$_4$). These have high stopping powers. But, because they are made from materials containing iodine, caesium, cadmium, and tungsten, they have absorption edges (Sections 3.1.2 and 5.5). This causes a dip in their efficiency in the region of the absorption edges. Count rates that can be detected are typically less that 10^5 cps, although some systems can operate up to 10^6 cps.

4.4. Solid state detectors

Solid state detectors for X-rays are made from either silicon doped with lithium (SiLi) or high-purity germanium (HPGE). Like scintillation detectors, the incident photon causes ionization, and the amount of ionization is proportional to the incident photon energy. The cloud of ionization produced is drawn to the anode of what is effectively a semiconductor diode. Thereafter, the current pulse is amplified and passes into a multichannel analyser system. Solid state detectors usually have reasonable energy resolution (150 eV for the SiLi and 180 eV for the HPGE). The operational energy range for these commences at 4 keV (determined by the thickness of the protective beryllium window). The upper limit for SiLi detectors is 20 keV, whereas the upper limit for HPGE is several MeV. The linewidth increases slightly with incident beam energy. The HPGE detector efficiency is not 100% over this range because of the L-edge absorption of germanium, which occurs around 11 keV (Fig. 6i). Both types of detectors are relative slow-speed devices ($<2 \times 10^4$ cps). When high count rates are required, multi-element detectors are used.

Solid state detectors find application in a wide range of experiments at synchrotron radiation sources. They are used for:

- X-ray fluorescence analysis of materials irradiated by the X-ray beam;
- energy-dispersive X-ray diffraction from small samples in high-pressure cells; and
- fluorescent XAFS and XANES experiments.

4.5. CCD array cameras

In Chapter 2, imaging techniques are discussed. The types of CCD array cameras that exist are too numerous to mention here. They are used variously for:

- the alignment of experiments, and
- Laue X-ray diffraction by small single crystals.

See, for example, the Photonic Science website: http://www.photonic-science.co.uk/.

4.6. Imaging plates

An imaging plate is a flexible plastic plate on which very small crystals (5 μm grain size) of a photo-stimulable phosphor is coated. This is usually $BAF(Br, I):Eu^+$. The phosphor stores the charge in proportion to the absorbed X-ray energy. When later stimulated by a laser, it releases the stored charge as photoluminescence. The intensity of this light is directly related to the absorbed energy from the incident photon. Imaging plates have a dynamic energy range of more than 10^5. Imaging plates are used widely for:

- alignment of equipment,
- X-ray powder diffraction studies, and
- small-angle X-ray scattering, where rapid time dependence of data is unimportant.

See, for example, the Fiji company website: http://www.fujimed.com/products-services/imaging-systems/.

4.7. Experimental hutches

Because the radiation emitted by the beams is penetrating and poses a risk to human health, shielding is necessary to protect the researchers. The recommended exposure limits (see Creagh and Martinez-Carrera, 2004) for radiation workers is 20 mSv per year. In general, experiments that operate with X-rays of energy above 3 keV have to be enclosed in a properly constructed lead-lined hutch. The experiments performed therein must be controlled remotely.

5. TECHNIQUES

It must be stressed that, for almost all studies of objects of cultural heritage significance, the use of many analytical techniques is necessary if definitive results are required (Creagh, 2005). What follows is a description of a number of synchrotron radiation techniques that the author has found useful in the study of these objects. I shall concentrate mainly on the techniques that are in common use (synchrotron radiation X-ray diffraction (SRXRD), synchrotron radiation X-ray fluorescence spectroscopy (SRXRF), XRR, GIXD, XAS, which incorporates XAFS and XANES) and the micro forms of these techniques (micro-SRXRD, micro-SRXRF, micro-XAFS, and micro-XANES). Also, I shall touch briefly on X-ray tomography. This topic has been discussed in Volume I of this book series by Casali (2006). Here, I will discuss questions of coherence and phase contrast, rather than absorption contrast techniques.

I intend to concentrate on the construction of the beamlines themselves rather than give a myriad of examples. Examples of the experiments performed with the equipment are to be found by accessing the website maintained by Pantos and Bertrand (http://srs.dl. ac.uk/arch/index.htm).

5.1. Synchrotron radiation X-ray diffraction (SRXRD)

The use of synchrotron radiation sources by materials scientists for cultural heritage studies is still comparatively new. However, the SRXRD technique has been used widely in these studies: studies that range from the study of pigments (Salvadó *et al.*, 2002) and corrosion products (De Ryck *et al.*, 2003) to fibres from Middle Eastern burial sites (Muller *et al.*, 2004). Dr. Manolis Pantos (http://srs.dl.ac.uk/people/pantos/), who is to contribute to Volume III of this book series, has been extensively involved in these studies.

5.1.1. "White beam" synchrotron radiation X-ray diffraction

In some experiments, "white" radiation is used, that is: no monochromators are used. The technique is sometimes referred to as "energy-dispersive" X-ray diffraction (EDXRD or SREDXRD) if synchrotron radiation is used. The dimensions of the synchrotron beam are set by slit systems, and the "pink beam" is allowed to fall on the sample. The synchrotron beam contains energy contributions in the range of 4 keV (determined by absorption in the beryllium windows that form the interface between the beamline vacuum and air) and $4E_c$, at which the intensity of a bending-magnet or wiggler source becomes low. The photons are detected by a single element germanium solid state detector placed at an angle of up to 30° to direct beam, and the pulse height distribution is determined using energy dispersive analysis techniques. Figure 7(a) shows a schematic figure of a typical "white beam" SRXRD system, in this case one in operation at Spring8 (http://www.spring8.or.jp/pdf/en/blhb/bl28b2.pdf). For EDXRD, the Bragg equation is used in its alternative form

$$2d_{hkl} \sin \theta = \left(\frac{hc}{E} \right)$$

where h is Planck's constant and E is the photon energy. Often, $2\theta = 30°$ is used for the detector setting, whence the interplanar spacings are given by

$$d_{hkl} = 1.932 \left(\frac{hc}{E} \right)$$

For further information, see, for example, Buras *et al.* (1994). The technique is further discussed by Bertrand later in this volume.

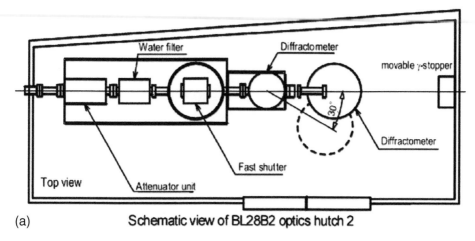

(a)   Schematic view of BL28B2 optics hutch 2

Fig. 7. (a) Schematic representation of a "white beam" X-ray diffraction device installed at Spring8. The lower (dotted) circle shows the location of the solid state detector when it is in use (http://www.spring8.or.jp/pdf/en/blhb/bl28b2.pdf). Note that a slit system has to be placed between the detector and the sample. The resolution of the system is determined in part by width of this slit and the $\Delta E/D$ of the solid state detector, which is typically 2% (say 150 eV at 7.5 keV).

One of the problems of using this technique is the fact, as mentioned in Section 4, that solid state detectors are easily saturated by the scattered beam intensities. The beam usually has to be attenuated to compensate for this. A benefit is that X-ray fluorescence from the specimen can be measured. Thus, both crystallographic and X-ray data are acquired for the same part of a specimen at the same time. Using small beams, microdiffraction and fluorescence spectroscopy can be accomplished at the same time. Creagh and Ashton (1998) have used EDXRD and XRF to study the metallurgical structure and composition of important and unusual medals, such as the Victoria Cross and the Lusitania medal. They showed that the German originals of the rare Lusitania medal, which were copied in the tens-of-thousands by the British (both nominally steels), were different in both lattice parameter and atomic composition.

5.1.2. Monochromatic synchrotron radiation X-ray diffraction using imaging plates

The most commonly used system of this type is at the Australian National Beamline (ANBF/PF) at the Photon Factory (referred to as BIGDIFF). This is essentially a large (573 mm radius) Debye–Scherrer or Weissenberg vacuum diffractometer using imaging plates to detect the diffracted radiation. In Fig. 7(b), a schematic representation is given of the use of the BIGDIFF diffractometer (573 mm radius) in its imaging plate mode, using its Weissenberg screen. The double-crystal monochromator is fitted with either [111] or [311] silicon crystals, and detuning is an option to remove higher harmonics. Sagittal focussing

b(i)

Fig. 7. (b) (i) Schematic representation of the use of the BIGDIFF diffractometer (573 mm radius) in its imaging plate mode using its Weissenberg screen. The IP cassette translated across the incident beam by a linear stage driven by a stepper motor. Also shown schematically is the double-crystal, sagittal-focussing monochromator.

of the upper crystal can increase the flux delivered to the sample by a factor of ≈ 10. The cassette, which is translated across the incident beam by a linear stage driven by a stepper motor, can be loaded with eight imaging plates, providing 320° coverage of reciprocal space.

This system has many uses, but its most common use is as a conventional powder diffractometer. That is, finely powdered samples are placed in capillaries that are spun about their axis during exposure to minimize preferred orientation and grain-size effects. X-ray diffraction is used to determine the crystal structures and phase compositions of materials. In conventional XRD experiments, only one capillary can be mounted and measured at a time. In contrast, the experiments at the (ANBF/PF) made use of an eight-position capillary spinning stage (Creagh *et al.*, 1998) in the vacuum diffractometer, operating with its Weisenberg screen. Seven specimens and one standard can be measured on

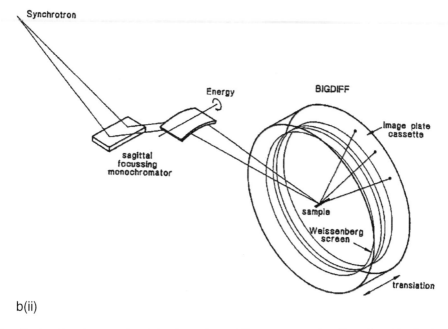

b(ii)

Fig. 7. (b) (ii) Photograph of the BIGDIFF diffractometer. The silver band is the inner surface of the imaging plate cassette, behind which is the outer case of the vacuum diffractometer. The copper arcs that can be seen are clamps for holding the imaging plates in position. At the centre of the diffractometer is a Huber θ–2θ goniometer, with angular encoders on each axis. BIGDIFF can be operated in the standard diffractometer mode when required.

one set of imaging plates. Exposure times are usually no more 10 min per sample. In addition, measurements can be made on less than 50 µg of material.

5.1.3. Scintillation radiation detector systems

Many types of diffractometers using scintillation counters and scintillation counter arrays can be used. In its simplest form, the system consists of a synchrotron radiation beam that has been rendered monochromatic by upstream monochromators, which impinges on a sample placed on a goniometer mounted on a conventional θ–2θ X-ray diffractometer. In the laboratory, these systems use Bragg–Brentano configuration: that is, the specimen rotates through an angle θ as the detector rotates through an angle 2θ. The specimen may or may not be moved around the θ axis. In Fig. 7(c)(i), the specimen remains still but the source and the detector are rotated clockwise and counterclockwise, respectively, to preserve the condition that the angle of incidence on the sample is equal to the angle of reflection from the sample. The angular position of the diffracted beam is measured using a high-speed detector mounted on the 2θ arm. Scintillator detectors are not as sensitive to high count rates as solid state detectors, but they can be saturated, and so care has to be

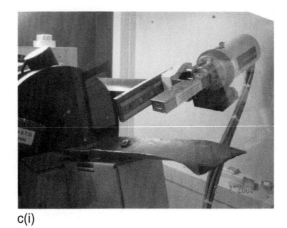

c(i)

c(ii)

Fig. 7. (c) (i) Schematic diagram for a conventional X-ray diffractometer. This is a θ–θ diffractometer, in which the sealed X-ray tube moves clockwise about the goniometer axis, and the solid state (HgCdTe) detector moves counterclockwise about the same axis. These motions can be linked to maintain the traditional θ–2θ motion associated with Bragg–Brentano geometry. In this case, the source is fixed at an angle close to the angle of total external reflection of the surface, and the detector is scanned to accumulate a diffraction pattern. In the experiment, the angle of incidence is varied, so that the depth of penetration of the X-rays can be varied, and the effect of surface coatings and surface treatment can be determined. The specimen here is part of the armour of a notorious Australian bushranger, Joe Byrnes (Creagh *et al.*, 2004). (c) (ii) Schematic diagram for the X-ray diffractometers at the Swiss Light Source. The sample (and its environmental chamber, when used) is carried on an Eulerian cradle and is located on the axis of the 2θ rotational stage. For the acquisition of low-resolution data, a microstrip detector is used. This detector is especially useful for data from systems in which the sample is changing with time; for example, phase changes in metals, and electrochemical changes at surfaces. For high-resolution data, a special diffracted-beam monochromator is used. This system has four monochromators, which are scanned around the 2θ axis to acquire the diffraction pattern.

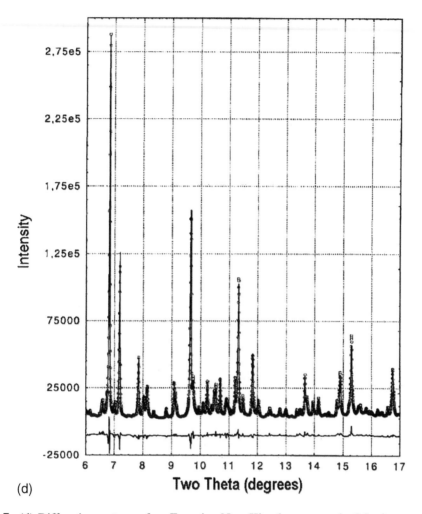

(d)

Fig. 7. (d) Diffraction pattern of an Egyptian New Kingdom cosmetic (Martinetto *et al.*, 2000). The observed data are the dots, and the solid line is the calculated diffraction pattern, calculated making allowance for preferred orientation, crystallite perfection, grain size, and styrin broadening. The curve given below is the difference between the observed and the calculated data.

taken in their use. Martinetto *et al.* (2000) used simple θ–2θ geometry in their study of Egyptian cosmetics at the European Synchrotron Radiation Facility (ESRF).

To reduce background scattering, diffracted-beam monochromators, tuned to the energy of the primary beam, are often used. Because the measurements are made sequentially, the time taken to acquire a spectrum can be long, especially for small samples. To overcome this, retaining the accuracy required for Rietveld analysis of the sample, a number of detectors are used. Figure 7(c)(ii) is a configuration used at the Swiss Light Source. The sample

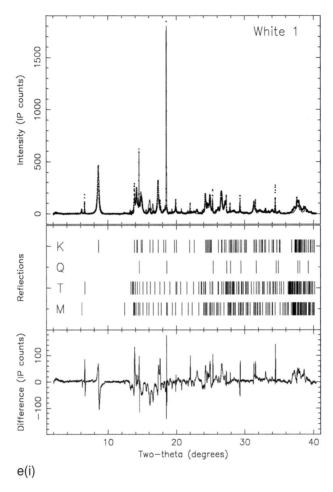

e(i)

Fig. 7. (e) (i) Synchrotron radiation X-ray diffraction pattern from a white pigment used by indigenous artists of the Arnhem Land region. The Rietveld fit to the experimental points is shown in the upper graph. Below that are the positions of the diffraction lines for kaolin (K), quartz (Q), talc (T), and muscovite (M). The Rietveld calculation is a whole-of-pattern fit to the experimental points. The bottom graph shows the difference between the observed and calculated values.

(and its environmental chamber, when used) is carried on an Eulerian cradle and located on the axis of the 2θ rotational stage. For the acquisition of low-resolution data, a microstrip detector is used. This detector is especially useful for data from systems in which the sample is changing with time; for example, phase changes in metals and electrochemical changes at surfaces. For high-resolution data, a special diffracted-beam monochromator is used. This system has four monochromators, which are scanned around the 2θ axis to acquire the diffraction pattern.

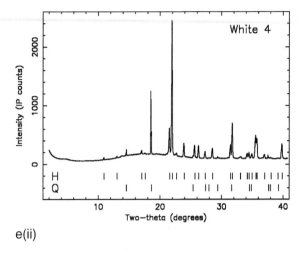

e(ii)

Fig. 7. (e) (ii) Synchrotron radiation X-ray diffraction pattern from a white pigment used by indigenous artists of the Arnhem Land region. The Rietveld fit to the experimental points is shown in the upper graph. Below that are the positions of the diffraction lines for hundtite (H) and quartz (Q). The Rietveld calculation is a whole-of-pattern fit to the experimental points. The bottom graph shows the difference between the observed and calculated values.

5.1.4. The Rietveld technique

The analysis of powder diffraction data from both X-ray and neutron experiments uses the Rietveld method (Rietveld, 1967; Young, 1993). The specimens are assumed to be a mixture of crystalline phases, each phase contributing its own pattern to the overall pattern.

To determine the combination of phases present in the sample, the Bragg equation

$$2d_{hkl} \sin \theta = \lambda$$

is used to make a list of values of d_{hkl} corresponding to the observed peaks. Association of measured peak positions with calculated or observed positions of pure single-phase fingerprints can be made using database search–match routines. Once the phases are identified, the subsequent step is quantitative phase analysis, which assesses the amount of each phase in the sample material, either as volume or weight fraction, assuming that:

- each phase exhibits a unique set of diffraction peaks; and
- the intensities belonging to each phase fraction are proportional to the phase content in the mixture.

A full-pattern diffraction analysis can, in addition to the phase fraction determination, include the refinement of structure parameters of individual mineral phases, such as lattice parameters and/or atom positions in the unit cell.

The Rietveld method allows the refinement of phase-specific structure parameters along with experiment-specific profile parameters by fitting a calculated model pattern to the entire observed diffraction pattern using the least-squares algorithm, which minimizes the quantity:

$$D = \sum_i g_i (y_i^{obs} - y_i^{calc})^2$$

The summation index i runs over all observed intensities y_i^{obs}. The weights g_i are taken from the counting statistics. y_i^{calc} are the calculated model intensities defined by instrumental and structural parameters, the latter including weight fractions in a multiphase refinement. By refinement of reflection profile parameters, crystallite size and microstrain effects can be studied. The Rietveld routine calculates figures of merit that indicate the quality of the fit of the entire model pattern to the entire observed diffraction pattern. A meaningful criterion is the weighted profile R-value R_{wp}:

$$R_{wp} = \left\{ \frac{\sum_i g_i (y_i^{obs} - y_i^{calc})^2}{\sum_i g_i (y_i^{obs})^2} \right\}^{1/2}$$

which should converge to a minimum. There are a number of programs available, many of them in the public domain, which can be used for X-ray as well as for neutron diffraction data analysis, $e.g.$ the General Structure Analysis System (GSAS).

The quantitative-phase information is obtained assuming that the weight fraction, W_p, of the pth phase in a mixture is given by the normalized product:

$$W_p = \frac{S_p Z_p M_p V_p}{\sum_i S_i Z_i M_i V_i}$$

where S_p, M_p, Z_p, and V_p are the refined Rietveld scale factor, the mass of the formula unit ($e.g.$ SiO_2), the number of formula units per unit cell, and the unit cell volume, respectively, of that phase p. The summation in the denominator accounts for all crystalline phases. Thus, in the case where not all crystalline components can be identified or in the presence of amorphous phases, the Rietveld analysis yields relative phase fractions only with respect to the crystalline phases contained in the model. The main advantages of quantitative phase analysis by the Rietveld method are as follows:

- No internal standard is required.
- Crystal structure models are included explicitly. Structure parameters can be refined along with weight fractions of mineral phases.
- Overlapping peaks and even peak clusters are handled without difficulty.
- Preferred orientation of crystallites can be considered in the model.

It may happen that one or more phases have been identified using reflection positions and extinction rules (yielding space group and lattice parameters), but structure models may not fit the experimental data because, for instance, powder grains are not statistically distributed in the object or simply because there are no complete structure models available, as is the case for some clay minerals like illite and kaolinite.

The presence of amorphous phases in samples makes a little difficult the interpretation of scattering data using the Rietveld method. In general, in Rietveld analysis, the background is stripped from the overall spectrum mathematically. As yet, no easy method exists for the quantitative interpretation of diffraction data in which there is a strong amorphous background.

5.1.5. Some measurements of cultural heritage materials using synchrotron radiation X-ray diffraction (SRXRD)

5.1.5.1. Diffraction study of Egyptian cosmetics from the New Kingdom Era. Martinetto *et al.* (2000) have used SRXRD to study the mineral composition of Egyptian cosmetics dating from the New Kingdom. Synchrotron radiation techniques are used because not much material is available for analysis. Particular problems occur when using historical samples because the mixture may contain mineral phases of different sizes and states of perfection. Figure 7(d) is the diffraction pattern of the cosmetic. Because the samples contained lead, a highly absorbing component, short wavelengths (0.09620– 0.04134 nm) were used. The Rietveld analysis program must be used carefully to account for the possibility of preferred orientation, crystal perfection, and grain sizes. The observed data are the dots, and the solid line is the calculated diffraction pattern. The curve given below is the difference between the observed and the calculated data. Accurate mineral-phase analysis is possible, and to give an indication of the information that can be extracted from the data, for their sample #E20514, the mineral phase composition was:

PbS (73%); $PbCO_3$ (3%); $Pb_2Cl_2CO_3$ (9%); $PbOHCl$ (1%); $PbSO_4$ (6%); ZnS (6%); $ZnCO_3$ (2%).

5.1.5.2. Diffraction studies of Australian aboriginal pigments. A study has been made of the diffraction patterns from the white and ochre pigments from a number of sites used by Australian indigenous artists using traditional techniques. The motivation for this is the need to find techniques for establishing the provenance of objects in museum collections. There is a need to be able to compare the mineral phase and trace element compositions in paint flakes taken from objects, which limits the size of the sample to at most 50 µg.

The white pigments are from Arnhem Land, in the north of Australia. The diffraction patterns had many lines, sometimes on a strong amorphous background. Figures 7(e)(i) and (ii) are typical diffraction patterns (O'Neill *et al.*, 2004). The diffraction patterns were analysed for composition using Rietveld analysis. As can be seen in Fig. 7(e)(i), the Rietveld refinement is reasonably good: perhaps as good as can be expected for whole-pattern fitting to clay minerals. The compositions of these white pigments and the percentage compositions are shown in Table 4. The row shown as "other" includes as yet unidentified phases and the contribution of the amorphous scattered background. It is interesting here to note that hundtite, seen occasionally in Arnhem Land pigments, is found extensively in cave paintings and objects from the Kimberley region. There seem to be strong regional differences in the white pigments.

Table 4. Compositions of white pigments used by indigenous artists in Arnhem Land

Sample	Mineral	% Composition
1	Kaolinite	74
	Quartz	18
	Talc	2
	Muscovite	3
	Other	3
3	Talc	98
	Hydroxyapatite	2
	Other	
4	Hundtite	81
	Quartz	4
	Other	15
5	Talc	98
	Hydroxyapatite	2
	Other	

Creagh *et al.* (2006a) have recently conducted a feasibility study to establish the amounts of pigments required to be able to make good X-ray diffraction and PIXE measurements on ochres. Ochre is a prized material in Australian indigenous culture. There were a limited number of mine sites, and trading of ochres occurred between communities. A detailed study of material from these mine sites has been made (Smith *et al.*, 2007) with a view to linking these to artefacts in museum custody.

5.2. X-ray-reflectivity (XRR) and grazing incidence X-ray diffraction (GIXD)

The equipment necessary for the study of surfaces and interfaces using XRR and GIXD is identical to that required for synchrotron radiation X-ray diffraction. In these cases, extended surfaces are investigated, rather than capillary tubes or specimens mounted on goniometer heads. It is necessary to mount the sample so that the surface of the sample lies on the axis of rotation of the θ and 2θ axes of the diffractometer. A very narrow beam height is necessary to ensure that the footprint of the X-ray beam at the smallest angle chosen falls on the sample.

5.2.1. XRR

In XRR experiments, the incident beam, the normal to the sample surface, and the detector (usually a high-speed scintillation detector) all have to lie in the same plane, and the surface must not be rough. Also, the detector must have a wide dynamic range. In Fig. 6(b), it can be seen that the reflectivity falls rapidly after the angle of total external reflection is reached. The intensity decreases as θ^{-4}. But it is in this region that the information about the nature of the interface is contained. To obtain a continuous curve, the incident beam is attenuated by a known amount until the reflect beam intensity reaches a level within the dynamic range of the detector (typically 1 MHz).

These types of experiments can be used to study thin coatings on surfaces. Creagh *et al.* (1999) undertook XRR studies of metal surfaces (aluminium and zinc) that had been coated with sodium dodecyl sulphate layers. BIGDIFF at the Australian National Beamline Facility at the Photon Factory was used. This was undertaken as part of a program of research studying the efficacy of protective waxes on bronze sculptures (Otieno-Alego *et al.*, 1999). The surfaces were analysed during electrochemical impedance spectrometry tests, the samples being removed from the saline bath at different times in the test regime. By measuring the fringe spacing in reflectivity graphs such as Fig. 8(a), it is possible to infer the film thickness. In Fig. 8(a), XRR data from a 21.7-nm layer of sodium dodecyl sulphonate on an aluminium substrate is shown.

5.2.2. GIXD

One of the problems with XRR measurements is that all the measurements are made in one plane of reflection. However, the film may not be amorphous: it may have crystalline structure in the plane of the surface. If this is the case, the application of Bragg's law (or its alternative expression, the Laue equations (Warren, 1968)) shows that peaks of intensity should exist. Experiments were undertaken with BIGDIFF without its Weissenberg screens in position. The incident beam was set to half the angle of critical reflection. The diffraction pattern was recorded on an imaging plate (Fig. 8(b)). In this figure, the axes are expressed in terms of Q $(=2\pi \sin \theta/\lambda)$. The figure shows the diffraction pattern of an ordered stearate film on a metal substrate. Note the sharp diffraction peaks. The streaks between the peaks are indicative of some disorder in the packing of the stearate molecules normal to the plane of the film. This shows that a hexagonal packing exists in the plane of the film and that, in this case, the two-dimensional ordering was present over a large area of the film. Such strong ordering is unusual in deposited layers. Self-ordering of the alkyl chains does occur, but only over small distances. These are oriented at random angles to the incident beam, so that Debye rings are seen, rather than a single-crystal-like diffraction pattern.

Modern diffractometers are able to undertake GIXD studies of processes taking place at the electrodes of electrochemical cells (Hallam *et al.*, 1997) studying, for example, the role of petroleum sulphonate corrosion inhibitors in protecting metal surfaces. More recently, real-time studies of processes taking an electrode surface as the electrode is driven through a full electrochemical cycle have been undertaken (De Marco, 2003). In studies of the effect of biologically active components of trace elements and nutrients in marine waters on electrochemical sensors, De Marco was able to show that the use of artificial seawater led to dissolution of oxides and the formation of haematite (Fe_2O_3). See Fig. 8(c).

Recently, Leyssens *et al.* (2005) have studied simultaneous *in situ* time-resolved SRXRD and corrosion potential analyses to monitor the corrosion on copper. The use of microstrip detectors (Fig. 7(c)(ii)) enables the acquisition of all the diffraction patterns for a predetermined time at any time in a cycle. Information about the build-up of crystalline

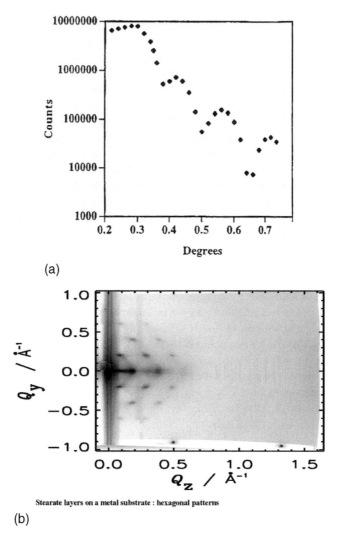

(a)

(b)

Fig. 8. (a) XRR data from a 21.7-nm layer of sodium dodecyl sulphonate on an aluminium substrate. (b) GIXD of an ordered stearate film on a metal substrate. Note the sharp diffraction peaks. The streaks between the peaks are indicative of some disorder in the packing of the stearate molecules normal to the plane of the film.

phases can be used in conjunction with synchrotron radiation FTIR data (Section 5.7) to derive very detailed information on electrochemical processes.

5.2.3. Small-angle X-ray scattering (SAXS)

Parts of both the XRR and GIXD data sets are taken in close proximity to the direct beam. In fact, the data shown in Figs. 8(a and b) were taken within 0.05° of the direct beam.

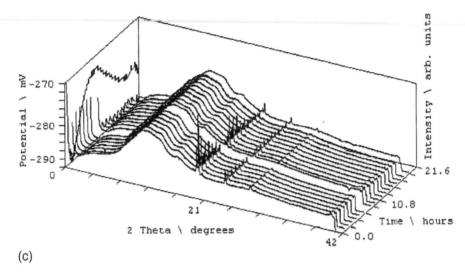

(c)

Fig. 8. (c) Typical time-resolved data in De Marco's time-resolved studies of the effect of artificial seawater on oxides. All the diffraction patterns are collected on the same imaging plate, positioning the cassette behind the Weissenberg screen for a given time, and then moving to another position after the desired sampling time has been reached.

Because data are taken close to the direct beam (to $Q = 0.005$ at 0.15 nm in some cases) SAXS beamlines must have extremely good collimation, good energy resolution, and small beam size. Detectors that are used can be area detector, such as the CCD MAR165 camera (MAR, 2006), or strip detectors. Devising a beamstop that will stop the direct beam, and yet be of sufficiently small size to enable the observation of low Q values, is a challenge.

Changing the length of the flight path between the detector and the sample changes the Q range. A schematic diagram of a possible arrangement is given in Fig. 10(e)(ii). For example, using a MAR 165 camera:

- at 1.5 Å, a Q range of 0.01–0.15 is achieved using a flight tube of length 8 m; and
- at 0.7 Å, a Q range of 0.4–1.8 is achieved for a flight tube length of 0.5 m.

Changing the flight tube is a cumbersome task because of their physical size.

It is not my intention to discuss the principles of SAXS in detail. These are discussed in some detail by Glatter and May (2004). Suffice here to say that SAXS can give both diffraction information and particle size and shape information, depending on the Q range chosen. For large Q values, diffraction patterns can be seen. The technique can be used in the study of fibre structure and morphology (Muller et al., 2004). For small Q values, values at which diffraction patterns cannot be generated, information is given on the size, habit, and morphology of crystals in materials. Wess et al. (2001) have used SAXS for studying the structural features of archaeological bones in which the habit of apatite crystals and recrystallized material may reflect the changes in bone environment since death.

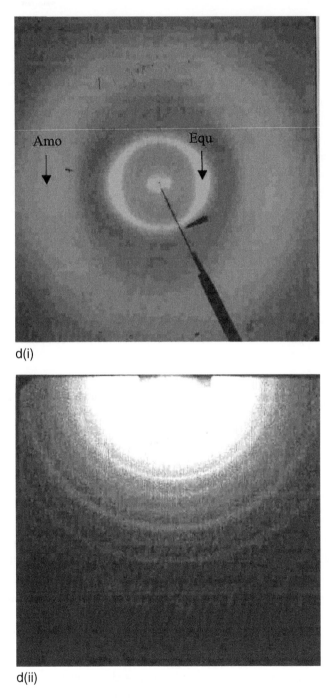

d(i)

d(ii)

Fig. 8. (d) (i) WAXD image of collagen. The equatorial reflection (Equ) arises from molecular interactions within a fibrin, and the outer ring (Amo) arises from amorphous arrangement of the polypeptide chains. (d) (ii) SAXS from skin. The strong diffraction rings represent the meridional series of collagen, because the electron density within the collagen is oriented axially along the fibre.

More recently, Kennedy and Wess (2006) have shown data taken at beamline ID18F at the European Synchrotron Radiation Facility (France) for both wide-angle and small-angle X-ray scattering from collagen taken from a historical parchment. Figure 8(d)(i) is a WAXD image of collagen. The equatorial reflection (Equ) arises from molecular interactions within a fibrin, and the outer ring arises from amorphous arrangement of the polypeptide chains. Figure 8(d)(ii) shows SAXS from skin. The strong diffraction rings represent the meridional series of collagen, because the electron density within the collagen is oriented axially along the fibre.

5.3. Microspectroscopy and microdiffraction

In microbeamlines, considerable attention is paid to the tailoring of the synchrotron radiation beam to provide a spot size that is ≈ 1 μm in diameter. The intention is to provide sub-micron resolution (≈ 0.1 μm) with the highest available flux for an energy range of 4–25 keV. These beamlines usually focus the beam on a sample mounted on an x–y–z translation stage, with xy motion providing the specimen positioning and the z-motion a measure of fine focussing. The same upstream geometry would be used whether the experiment were to be microdiffraction, microspectroscopy, micro-XAFS, or micro-XANES.

Small source size and source directionality are very important design requirements. In the Australian Synchrotron, the source will be a 22 mm undulator of 90 periods length, and the gap length should be 6 mm. The pole tip field is 0.83 T, and $K = 0.9337 \lambda_u B_o = 1.71$. In this case, the photon energy of the fundamental is

$$E_1 = \left(\frac{0.95E^2}{1+(K^2/2)} \right) \lambda_u = 1.583 \text{ keV}$$

Also, a spectral purity ($\Delta E/E = 10^{-4}$) is required. With these specifications, 1 ppm detection limits ought to be achieved.

Figure 9(a) is a schematic diagram of the microfocus undulator beamline proposed for the Australian Synchrotron (Paterson et al., 2006). The whole beamline is operated at UHV levels. Radiation from a 22-mm period, 2000-mm-long undulator with $K = 1.478$ passes through the shield wall to a slit system situated at 13.25 m from the source. The radiation falls on a double-crystal silicon monochromator situated at 14.5 m from the source. Cooled crystals are required, and two sets of [111] and [311] silicon crystals will be required to enable operation of the system in the range 4.7–25 keV. The divergent beam from the monochromator must be focussed in the vertical and horizontal planes onto the sample, or further optical elements. This is performed using two mirrors, one horizontally focussing, and the other vertically focussing. Figure 9(a)(ii) is a schematic diagram showing how two curved optical elements can be used to produce a focussed beam. This configuration is referred to as a Kirkpatrick–Baez pair. The minimum angle of incidence of the central ray on the mirrors is 2 mrad. Three strips of metallic coating are placed on the mirrors to aid in harmonic rejection. These are rhodium, platinum, and silicon. The choice

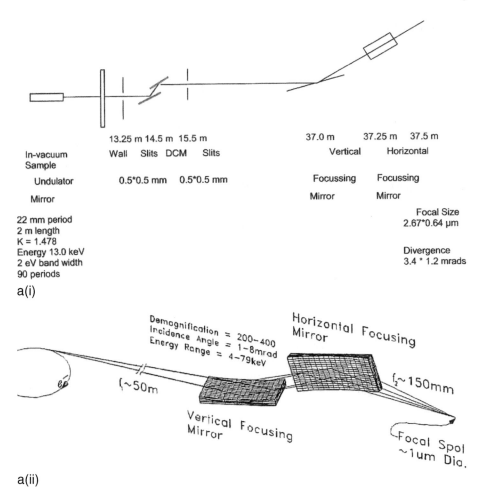

a(i)

a(ii)

Fig. 9. (a) (i) Schematic diagram of the configuration of optical elements in the microspectroscopy beamline at the Australian Synchrotron. (a) (ii) Schematic diagram showing how two curved optical elements can be used to produce a focussed beam. This configuration is referred to as a Kirkpatrick–Baez pair.

of which strip is driven into the beam is determined by the energy of the incident beam (Section 3.1.2). With the proposed system, the focal size is expected to be 2.67 μm × 0.64 μm, and the divergence is 3.4 mrad × 1.2 mrad at 13 keV. The flux in a 2-eV bandwidth is expected to be 1.19×10^{12} photons/s.

Smaller focal lengths can be achieved by the insertion of other optical elements. Fresnel zone plates (Section 3.3) can be used to produce smaller spot sizes. With these, submicron-focussed spot sizes can be achieved in the required energy range. The configuration of the microspectroscopy beamline when Fresnel lens focussing is used is given in Fig. 9(b).

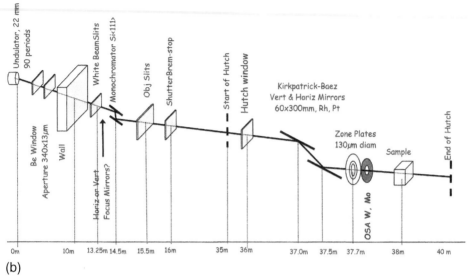

(b)

In-situ Identification of Micro-inclusions in Rhodonite

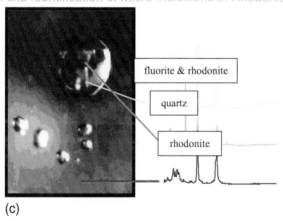

(c)

Fig. 9. (b) Schematic diagram of the microspectroscopy beamline at the Australian Synchrotron when Fresnel lens focussing is used. (c) Inclusions in the semiprecious gemstone, rhodonite. The minerals can be identified by using Raman microscopy. But crystallographic diffraction techniques (XRD) and compositional measurements (XRD) are necessary to confirm the identification.

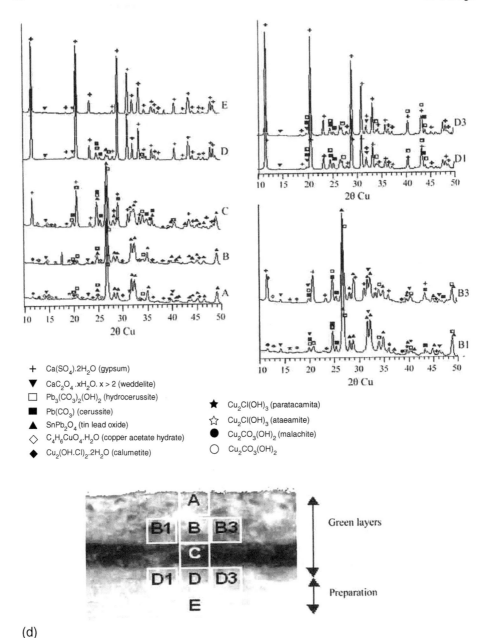

+ Ca(SO$_4$).2H$_2$O (gypsum)
▼ CaC$_2$O$_4$.xH$_2$O. x > 2 (weddelite)
□ Pb$_3$(CO$_3$)$_2$(OH)$_2$ (hydrocerussite)
■ Pb(CO$_3$) (cerussite)
▲ SnPb$_2$O$_4$ (tin lead oxide)
◇ C$_4$H$_6$CuO$_4$.H$_2$O (copper acetate hydrate)
◆ Cu$_2$(OH.Cl)$_2$.2H$_2$O (calumetite)

★ Cu$_2$Cl(OH)$_3$ (paratacamita)
☆ Cu$_2$Cl(OH)$_3$ (ataeamite)
● Cu$_2$CO$_3$(OH)$_2$ (malachite)
○ Cu$_2$CO$_3$(OH)$_2$

(d)

Fig. 9. (d) SRXRD patterns corresponding to a cross section of a sample of the altarpiece of Constable. The diffraction patterns were taken with a collimator or 100 μm diameter, using a single-bunch beam from station 9.6 at the SRS Daresbury. The beam current was 20 mA, and the exposure time was 30 s. The location of the diffraction pattern on the pain chip is indicated by letters. Data used by courtesy of Dr. Manolis Pantos.

With highly focussed, well-monochromated, intense beams available, what experimenters can do with the beam is limited only by their imagination. It is possible to study the diffraction by small crystallites and inclusions in metals, alloys, and minerals using thin foil or films, and geologically thin sections. The transmitted Laue patterns can be recorded on imaging plates or on CCD cameras. At the same time, it is possible to measure the composition of the object of interest. Using the automated stage, it would be possible to study minerals and organic materials trapped on micropore aerosol or stream filters.

The only constraints are on how best to present the samples to the beam.

In what follows, some applications to material of cultural heritage significance will be given.

5.3.1. Microdiffraction (micro-XRD)

It is possible to study the structure of single fibres of material, such as the cottons, flaxes, and wools found in ancient tombs and other archaeological sites. An example of this is the work of Muller *et al.* (2000), who studied the small-angle scattering and fibre diffraction produced on single-cellulose fibres, and Muller *et al.* (2004), who studied the identification of ancient textile fibres from Khirbet Qumran caves.

The identification of inclusions in geologically thin sections is possible. It should be noted that microdiffraction is not the only technique that can be used: Raman microscopy can be used for mineral-phase identifications. Figure 9(c) shows an optical micrograph of inclusions in the semi-precious gemstone, rhodonite. Identification of the mineral phases depends on micro-XRD experiments. Quartz and fluorite may be found in the inclusions (Milsteed *et al.*, 2005).

Salvado *et al.* (2002) utilized beamline 9.6 at the SRS Daresbury laboratory, operating in single-bunch mode (2 GeV, 20 mA). A very small (100 μm diameter) X-ray beam was used to scan across the cross section of a sample. The sample was a paint chip embedded in a plastic resin. Transmission geometry was used with the XRD patterns being collected by a QUANTUM-4 CCD area detector. Figure 9(d) shows SRXRD patterns corresponding to a cross section of a sample of the altarpiece of Constable.

5.3.2. Microspectroscopy (micro-SRXRF)

As mentioned earlier, the role of the beamline is to deliver a well-conditioned, finely focussed beam to the specimen. The options for those who seek to undertake microscopy experiments are many, and are limited largely by the size of the specimen translation stage, the detectors available, and how they can be located within the experimental hutch. In studies of corroded ancient bronze objects, De Ryck *et al.* (2003) used a wide range of experimental techniques to establish which was the most appropriate. In this study, both SRXRD and SRXRF were performed.

Recently, at BESSY II, Guerra *et al.* (2005) has continued her research on coinage by fingerprinting ancient gold by measuring Pt with spatially resolved high-energy SRXRF.

5.4. XAS

A review of XAS has been given by Creagh (2004a). An example of a typical XAS beamline is shown in Fig. 6(a). Figure 10(a)(i) shows an artist's impression of the XAS beamline at

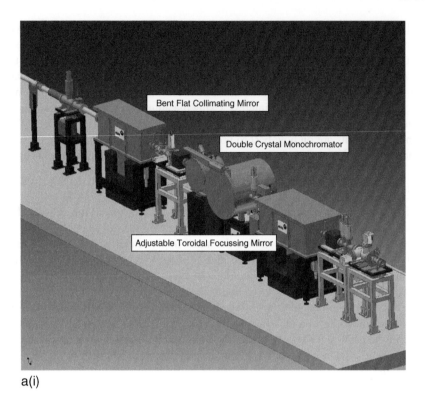

a(i)

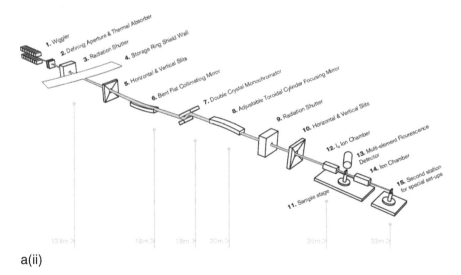

a(ii)

Fig. 10. (a) (i) Artist's impression of the XAS beamline at the Australian Synchrotron. This will be operated in conjunction with a wiggler source (the details of the front end linking to the storage ring is not shown), which will extend the upper energy significantly. The locations of the vertical and horizontal focussing mirrors are shown, as is the double-crystal monochromator. (a) (ii) The disposition of optical elements in the beamline shown in Fig. 10(a)(i).

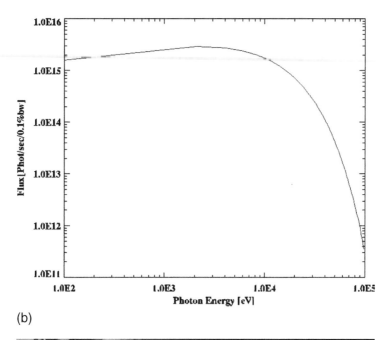

(b)

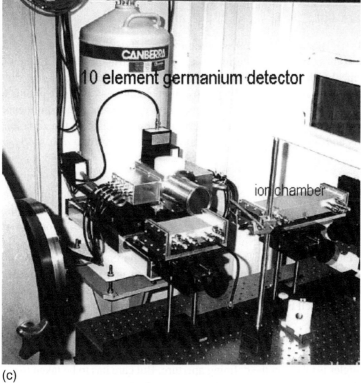

(c)

Fig. 10. (b) Calculations of the photon flux of the wiggler as a function of photon energy. (c) The XAFS station at the Australian National Beamline Facility at the Photon Factory. Shown are the ionization chambers for transmission XAFS experiments and the 10-element germanium detector. This has now been superseded by a 30-element detector.

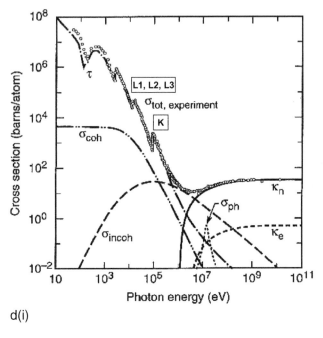

d(i)

Fig. 10. (d) (i) The absorption cross section of molybdenum is plotted as a function of energy. The contributions of scattering by the photoelectric, Rayleigh, Compton, and pair production mechanisms are shown. Each element has a different scattering cross section. In the energy region for which XAFS is a useful technique (4–40 keV), photoelectric scattering is the dominant mode. The circles represent experimental measurements. Note the discontinuities on the curve for total scattering. These correspond to excitation levels for atomic shells and subshells, and the discontinuities are referred to *absorption edges*.

the Australian Synchrotron (Glover *et al.*, 2006). This will be operated in conjunction with a wiggler source (the details of the front end linking to the storage ring is not shown in this figure), which will extend the upper energy significantly (Fig. 10(b)). For the W61 wiggler, the flux is nearly 10^{16} photons/cm^2 at 0.1% bandwidth at around 10 keV. The flux remains above 10^{12}, close to 100 keV. The increase in flux and energy range is significant. The overall range is such that excitation of K-shell electrons up to the uranic elements is possible. And the flux is very high in the region of 10 keV: close to the K-shell of the transition metals, and the L-shell emissions of all the elements up to uranium.

The locations of the vertical and horizontal focussing mirrors are shown in Fig. 10(a)(ii), as is the double-crystal monochromator. The wiggler has magnets of ≈ 1.8 T field strength, K is ≈ 20, and there are 18 periods with a period length of 110 mm. Some of the heat load from the wiggler is taken by the defining aperture. But care has to be taken to cool the bent flat vertically focussing mirror. Also, the mirror performs a harmonic rejection role, and is strip coated with silicon, rhodium, and platinum to ensure that harmonic content is a minimum. These have energy cut-offs of 13, 23, 34 keV, respectively. Above 34 keV, the mirror

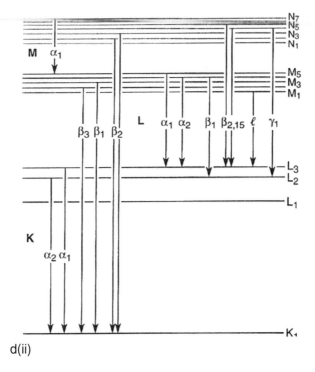

d(ii)

Fig. 10. (d) (ii) A schematic diagram of energy levels in an atom showing how different photons produce radiations of different energies when transitions occur between levels. Transitions to the innermost electron shell of an atom (the K-shell) from the next shell (the L-shell) gives rise to photons with slightly different energies (referred to as $K_{\alpha 1}$ and $K_{\alpha 2}$). If an electron is removed from the K-shell by the absorption of a photon, the electrons within the atom rearrange themselves, and a cascade of electronic transitions occur, causing the emission of photons of various discrete energies (LI, LII, LIII; MI, MII, MIII, MIV, MV, etc.).

will be moved out of the beam. Further beam definition will be performed using beam-defining slits.

The double-crystal monochromator has to have good energy resolution ($\Delta E/E < 10^{-4}$) and two cuts of silicon monochromator crystal have to be available: [111] for the range 4–24 keV; [311] for the range 4–50 keV. The first of each crystal pair has to be cryogenically cooled to reduce the effect that thermal loading has on the reflection of the X-rays. Because, as we shall see later in this section, the X-ray energy is scanned during an experiment, the exit beam from the monochromator should remain the same, whatever energy is chosen. Reproducibility of position is also a strong requirement, since, when XANES spectra are taken, energy differences of 0.1 eV in a measurement of 6.4 keV can be important. If time-resolved studies are to be undertaken, the time taken for a scan becomes a significant parameter in experimental design. The monochromator has to be able to be operated in the QEXAFS mode (Bornebusch *et al.*, 1999).

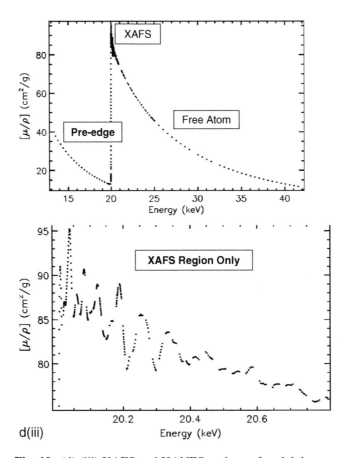

Fig. 10. (d) (iii) XAFS and XANES regions of molybdenum.

The adjustable toroidal mirror provides both vertical and horizontal focussing. The aim is to provide a beam of ≈(0.2 mm × 1 mm) to the sample area. This mirror is rhodium coated, and it can be rotated about a horizontal axis. At above 23 keV incident beam energy, this mirror is removed from the system. Since both focussing elements are removed from the system above 23 keV (just above the K-shell absorption edge of molybdenum), the beam size is determined by the slit apertures.

The end station hutch contains the experimental apparatus required for XAS studies. Ion chambers are used extensively in XAS experiments, both to monitor the beam intensity and to measure the transmitted X-ray beam. XAS experiments use either transmission beam measurements or the fluorescence radiation. For the latter, multiple element (up to 100) germanium solid state detectors are used. Figure 10(c) shows the XAFS experimental station at the Australian National Beamline Facility (ANBF), the Photon Factory, Tsukuba, Japan. The sample is situated between two ionization chambers. A 30-element germanium detector is located at right angles to the specimen surface for fluorescence XAFS experiments.

A wide range of sample stages and mounts are used. These may include cooling and heating stages, and be able to operate with liquids and solids, or at high pressures. Cryogenic cooling of the sample is desirable to reduce the effect of thermal motion of atoms in the sample, where this is consistent with experimental constraints. This can sometimes be used to the advantage of the researcher because chemical reactions taking place outside of the sample cell can be transported into the sample cell, and "stop-frozen" to give information of the reaction products present.

5.4.1. X-ray absorption fine structure (XAFS)

XAFS useful in probing the environments of selected atomic species in a material. As a technique, it is used by biologists, chemists, physicists, materials scientists, engineers, and so on, for monitoring changes in systems, sometimes statically and sometimes dynamically. Since XAFS is a less familiar technique to conservation scientists, I shall give a brief description of it and how it is used to probe the behaviour of materials. The discussion will centre on transmission measurements. Fluorescence radiation is related to transmission radiation through the fluorescence yield, a tabulated function. See http://www.csrri.iit.edu/periodic-table.html (Krause, 1979).

As its name suggests, XAFS has its origin in the measurement of the transmission photons through materials. The absorption of atoms has been tabulated by Creagh (2004b) for all atoms from atomic numbers 1–92. If the intensity of a photon beam is measured before and after its passage through a material of thickness (t), the relation between the incident beam intensity (I_o) and the transmitted intensity (I) is given by

$$I = I_o \exp(-\mu_1 t)$$

Here, μ_1 is the linear attenuation coefficient. This is related to the better known mass absorption coefficient by

$$\mu_m = \left(\frac{\mu_1}{\rho} \right)$$

A number of different processes contribute to the absorption coefficients: photoelectric absorption, Rayleigh scattering, Compton scattering, and pair production. In Fig. 10(d)(i) the absorption cross section of copper is plotted as a function of energy. The contributions of scattering by the photoelectric, Rayleigh, Compton, and pair production mechanisms are shown. Each element has a different scattering cross section. In the energy region for which XAFS is a useful technique (4–40 keV), photoelectric scattering is the dominant mode. The circles represent experimental measurements. Note the discontinuities on the curve for total scattering. These correspond to excitation levels for atomic shells and sub-shells, and the discontinuities are referred to absorption edges.

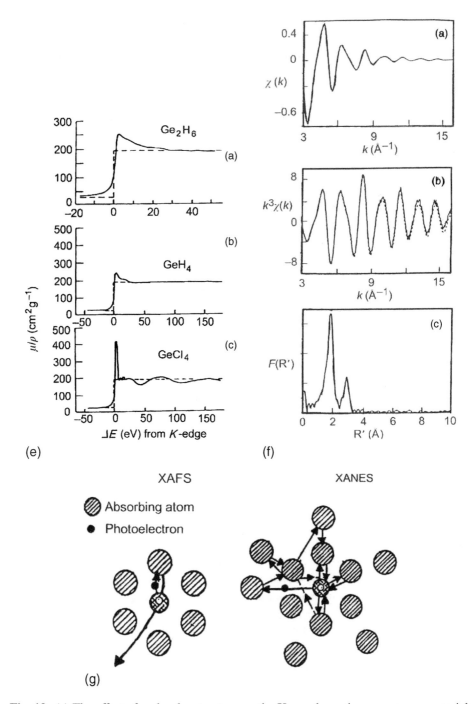

Fig. 10. (e) The effect of molecular structure on the X-ray absorption on gaseous materials in the region of the K-shall absorption edge: GeCl$_4$, GeH$_4$, Ge$_2$H$_6$. Note that these spectra are XANES spectra since there can be no long-range ordering in the gas phase. (f) Steps in the XAFS data reduction process. (g) Illustration of the distinction between XAFS and XANES.

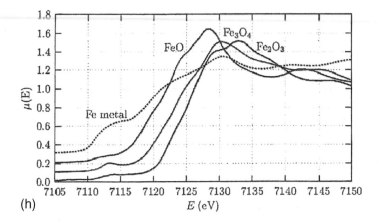

(h)

Fig.10. (h) XANES and pre-edge structures in iron oxides (Foran, 2005). A knowledge of these structures is, for example, essential for the understanding of the degradation of historic iron-gall inks on parchment.

Figure 10(d)(ii) is a schematic diagram showing how photons of different energies are produced when transitions occur between levels. For XAFS studies, absorption by the K-, L-, and M-shells of atoms are of interest.

The calculations given in tabulations such as Creagh (2004c) are "single atom" calculations: that is, they are made for a single isolated atom. Using these values to calculate the absorption of an assemblage of atoms works well at energies remote from an absorption edge. However, as the absorption edge is approached, deviations from the free atom model occur. Dramatic effects can occur. Figure 10(d)(iii) shows the mass attenuation coefficient taken for an energy region that includes the K-shell absorption edge. The upper figure identifies the pre-edge XAFS and free-atom regions. The lower figure shows just the XAFS data.

Figure 10(e) shows the effect at the edge for three germanium compounds. Evidently the local environment of the target atomic species, germanium, strongly influences the measured absorption coefficient.

It is this fact that has lead to the use of XAFS by scientists to probe the local environments in atoms, and changes that may occur in, say, chemical reactions. The experimental geometry used in XAFS measurements does not allow absolute measurements to be made. Readers should peruse the articles of Chantler's group (2002–2006) to understand the difficulties of making absolute XAFS measurements. Specimen thickness can have a smearing effect on the XAFS spectra. Thickness should be in the range $2 < \ln(I_0/I) < 4$ for best results.

Nevertheless, XAFS is a valuable research tool, albeit a relative one.

Before giving a brief description of the theory underlying XAFS, the distinction between XAFS and XANES should be made. XAFS occurs over a considerable energy range before the edge, 1 keV or more. XANES occur in the energy range, say 100 eV before and 100 eV after the absorption edge. The origins of the two effects are different, as shall be explained later. Both effects give information on: the oxidation state of the target atom, its local coordination geometry, the type and number of its neighbouring atoms, its bond lengths, the angle between bonds, and molecular orientation.

XAFS is an interference effect caused by the interaction of the ejected photoelectron with its surroundings. The first step in XAFS theory is to determine the value of $\chi(E)$ at each energy step in the XAFS pattern.

$$\chi(E) = \frac{\mu_1(E) - \mu_{10}(E)}{\mu_{10}(E)}$$

where $\mu_{10}(E)$ is the extrapolation of the "pre-edge" or "free atom" region to the edge energy. This is done by spline fitting the data.

As mentioned, XAFS is an interference effect caused by the interaction of the ejected photoelectron when a photon ($\lambda = 2\pi/k$) is absorbed. Here:

$$k = \left(\frac{m}{2\pi^2}(E - E_0) \right)^{0.5}$$

It is convenient to describe the $\chi(E)$ in terms of the photoelectron's momentum, k.

$$\chi(k) = \Sigma (N_i k r_j^2) |f_i(k)| \exp\left(\sigma_i^2 k^2 - \frac{r_i}{\rho} \right) \sin[2kr_i + \varphi_i(k)]$$

where the summation occurs over the shells of atoms that surround the target atom; N_i is the number of atoms in the ith shell; r_i is the distance of the shell from the target atom; $f_i(k)$ is the scattering amplitude to which is associated a phase-shift $\varphi_i(k)$; $f_i(k)$ depends only on the backscattering atom, whereas the phase depends on contributions from both the scattering and the absorbing atom; $\varphi'_{i'}(k) = \varphi'_{j'}(k) + \varphi'_{i'}(k) - l\pi$, where $l = 1$ for K and L1 edges and 2 or 0 for L2 and L3 edges; and $\exp(\sigma_i^2 k^2)$ is related to harmonic vibrations (referred to as the Debye–Waller factor).

The data reduction procedure is extensive for XAFS data. However, the maturity of XAFS data reduction programs ensures that this causes a minimum of discomfort to the researcher. All major research facilities will maintain these programs. For example, for the Australian Synchrotron Research Program's data reduction program XFIT, see http://www.ansto.gov.au/natfac/asrp7_xfit.html.

Steps in the data reduction process are:

- converting measured intensities to $\mu(E)$,
- subtracting a smooth pre-edge function,
- normalizing $\mu(E)$ to the range 0–1,
- fitting a curve to the post-edge $\mu(E)$ values to approximate $\mu_0(E)$,
- calculating $\chi(E)$, and
- identifying the threshold energy E_0 and converting to k-space to determine $\chi(k)$.

Having done this, the XAFS data is in a form from which information can be extracted. Figure 10(e) shows further steps in the data reduction process:

- $\chi(k)$ is weighted by a factor of k^3, so that the low-amplitude, high-energy part of the XAFS oscillations are increased in amplitude (referred to as $\chi(R)$);
- electronic filter window is placed around this to limit the range over which; and
- the Fourier transform of $\chi(R)$.

From this point on, a modelling process is undertaken. Various configurations of environments surrounding the target atom are considered and the information is used in the theoretical equation for $\chi(k)$. The data for the models may be based on crystallographic information; for example, if the material under investigation were FeO, the iron atom would be in an octahedral configuration with the oxygen atoms, and R_1, the radius of the first shell, would be 2.14 Å. The modelling proceeds until a match between experiment and theory is achieved.

Despite all the data manipulations that have to be made, and the fact that the analysis gives relative rather than absolute information, XAFS is a powerful analytical tool. Perhaps 60% of all experiments at the ANBF are XAFS experiments.

In one of several experiments that have been undertaken, Pantos et $al.$ (2002) have demonstrated the use of SRXRD and XAFS for the study of the mineral composition of painting pigments and pottery glazes.

5.4.2. X-ray absorption near edge structure (XANES)

In the case of XAFS, the photoelectron is ejected from the atom and is, therefore, transformed from a bound state to the continuum. The electron interacts with neighbouring atoms, the high-energy electrons undergoing single scatter, and the low-energy electrons undergoing multiple scatter in their passage away for the target atom (Fig. 10(g)). For XANES, the situation is different because the transitions may be to virtual energy states, or to other bound states. Therefore, XANES gives information about the coordination chemistry of the target atom.

Most experiments have been performed using focussed beams. The strong dependence of pre-edge and near-edge structure on ionization state is shown in Fig. 10(h) for the oxides of iron (Foran, 2005). These measurements are a precursor to the study of iron speciation on iron-gall inks on parchment (Lee et $al.$, 2006).

De Ryck et $al.$ (2003) have used XANES to study oxidation state maps involving the copper ions in corrosion products formed in a bronze object.

5.5. Infrared (IR) techniques

The use of the infrared (IR) component of the synchrotron spectral output for research is relatively recent (less than 10 years). This field of research, pioneered initially by Dr.Gwyn Williams at the NSLS (Williams, 1990), has led to the establishing of beamlines at a number of synchrotron radiation sources. Synchrotron radiation infrared radiation (SRIR) has applications to the fields of surface science, geology, cell biology, materials science, conservation science, and so on. Initially, this research was performed using synchrotron radiation from sources of less than 1 GeV energy, but more recently storage energies of up to 8 GeV have been used (ESRF, Spring8). IR beamlines are currently being designed for SOLEIL (2.75 GeV) and the Australian Synchrotron (3 GeV).

SRIR sources have a number of advantages over conventional IR (black body) sources. Table 1 sets out a comparison of the capabilities of radiation sources of various types. A conventional source radiates over a large area, and into all space. SRIR is up to 1000 times more intense over the whole wavelength range (10–100 µm) than a conventional source (Fig. 11(a)(i)). And it delivers a collimated beam from a source of small size (Fig. 11(a)(ii)). This shows the unapertured beam profile at the sample stage at the IR Beamline 11 at CLRC Daresbury (Tobin, 2006). The mapped area is 30×30 µm^2. The brightness of the synchrotron radiation source makes it especially useful in microscope/microprobe experiments.

5.5.1. Design requirements and constraints

The functional requirements of an IR beamline at the sample is for the provision of radiation in the wavelength range, typically 0.4–100 µm, with a photon flux of around 5×10^{13} photons/s/0.1% bandwidth. In newer facilities, it is a requirement that both the high-resolution spectrometer and the microscope–spectrometer be able to be used simultaneously by research groups.

IR beamlines differ significantly from other beamlines found at synchrotron radiation sources because the mechanism for the extraction of the radiation involves the positioning of the extracting mirror within the vacuum chamber of the synchrotron. This places severe constraints on the design of the system. The design requirements for all IR beamline systems are the:

- maximization of the cone of radiation extracted (which, in practice, is related to the design of the vacuum chamber);
- preservation of the integrity of the ultra-high vacuum within the ring; the minimization of vibration in the mirror system;
- need to control the heat load on the mirror due to the hard component of the synchrotron radiation beam; and
- need for accurate mirror positioning.

All this has to be achieved within the constraints posed by the bending magnet and associated beam-focussing magnets (Fig. 11(b)). The vacuum chamber had to be modified to enable the extraction mirror to be fitted. In almost all SRIR beamlines a design solution enabling vertical extraction of the beam is possible. However, in the case of the Australian Synchrotron, because the configuration of the vacuum chambers was optimized for X-ray production, the only modification to the chambers that would enable a suitably large extraction aperture involves the use of a horizontally mounted mirror. It is important to accept a large solid angle since the flux at long wavelengths is sensitive to the angle of acceptance.

Modifications were made to the dipole vacuum chamber to enable an extraction area of 17 mrad (vertical) and 58 mrad (horizontal) to be obtained. This is 40% larger than what might have been obtained using vertical extraction. This results in a gain in brightness at 100 mm of around 100%.

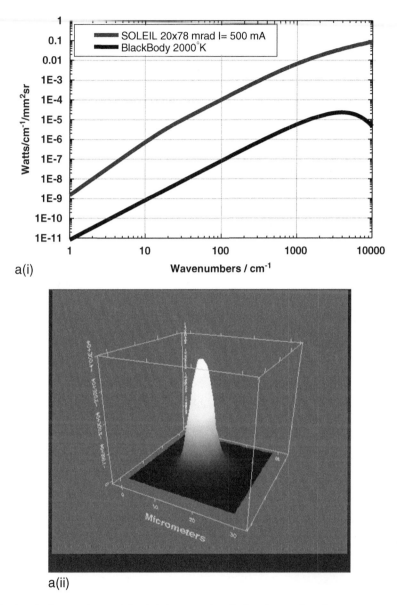

a(i)

a(ii)

Fig. 11. (a) (i) Comparison of the brightness of radiation from a bending-magnet synchrotron radiation source operating at 3 GeV, 200 mA, with that of a black body source operating at 2000 K. (a) (ii) The unapertured beam profile at the sample stage at the IR Beamline 11 at CLRC Daresbury (Tobin, 2006). The mapped area is 30×30 μm^2.

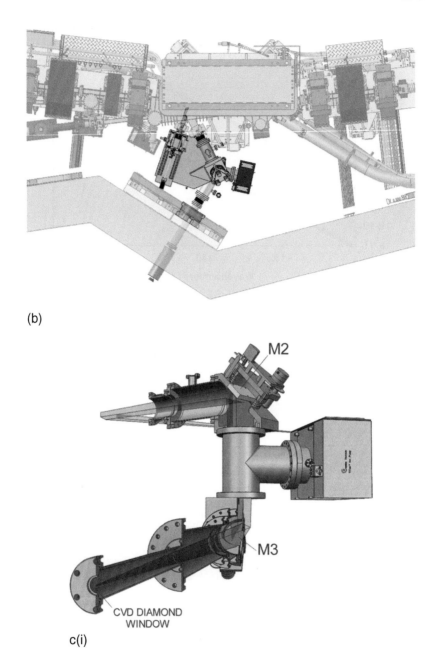

(b)

c(i)

Fig. 11. (b) Bending and focussing magnets surrounding the vacuum chamber. (c) (i) The beam path through the front end mirror system at the Australian Synchrotron.

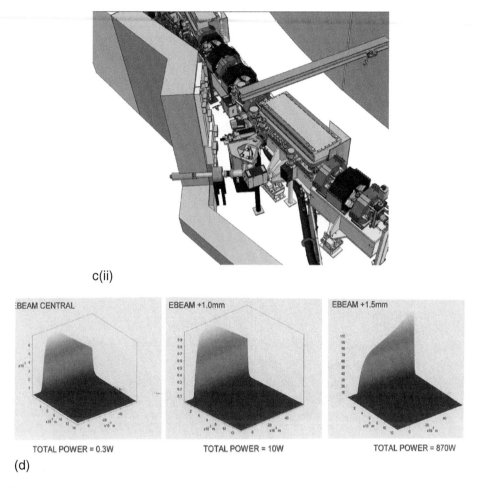

c(ii)

(d)

Fig. 11. (c) (ii) Another view of the beam path through the front end mirror system at the Australian Synchrotron, showing the location of ion pumps. (d) The effect on the heatloading on mirror M1 due to deviations from the true orbit. From left to right: true orbit, 1 mm deviation; 1.5 mm deviation.

The configuration of the beamline inside the radiation shield wall is shown in Figs. 11(c)(i) and (ii). Figure 11(d)(i) shows the beam path from the sources (one component arises from the so-called "edge radiation", and the other from the bending-magnet radiation) to the mirror M1, and from there to the ellipsoidal focussing mirror M2 and the plane deflecting mirror M3. In the final design, a mirror pair will be used to perform the function of M3. The IR radiation exits the UHV section of the beamline (1×10^{-9} mbar) through a CVD window. Figure 11(d)(ii) shows the ancillary beamline components such as the ion pumps and the gate valves used to isolate the extraction system from the vacuum in the storage ring.

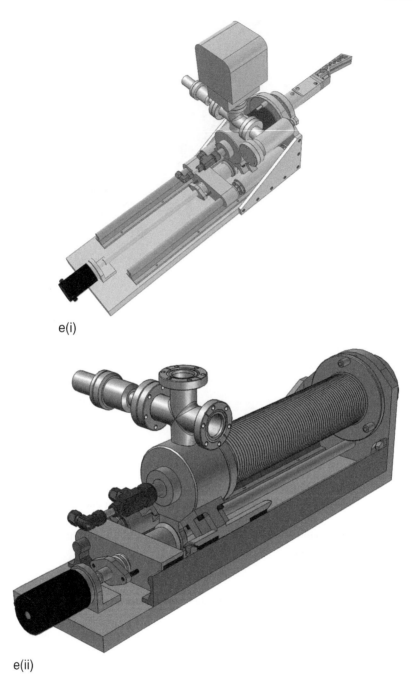

e(i)

e(ii)

Fig. 11. (e) (i) View of the mirror M1 in its inserted position.

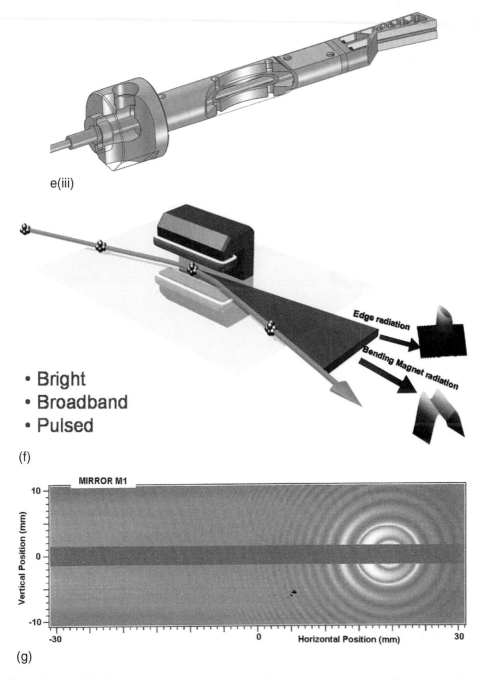

e(iii)

- **Bright**
- **Broadband**
- **Pulsed**

Edge radiation

Bending Magnet radiation

(f)

MIRROR M1

Vertical Position (mm)

10

0

-10

-30   0   Horizontal Position (mm)   30

(g)

Fig. 11. (e) (ii) Cutaway view of the proposed concentric water cooling system for the mirror M1. The machined Glidcop mirror is bolted onto the stainless steel mirror beam. (f) Illustrates the formation of bending-magnet and edge radiation. (g) Edge (circular fringes) and bending magnet (horizontal bands) at 3 μm, calculated using the SRW method.

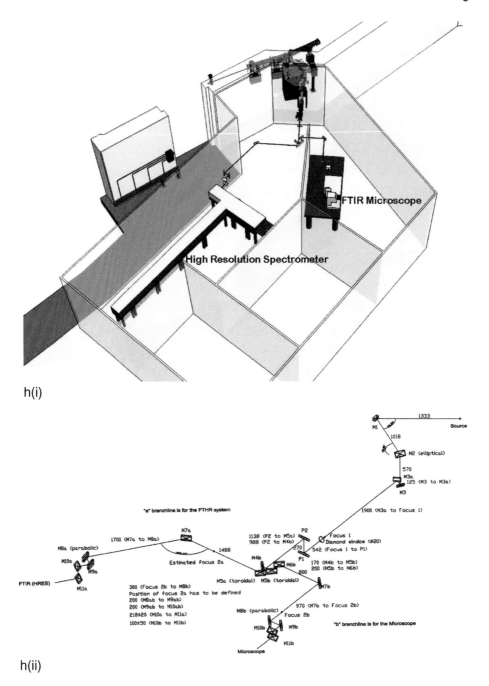

h(i)

h(ii)

Fig. 11. (h) Plan view of the IR beamline at the Australian Synchrotron.

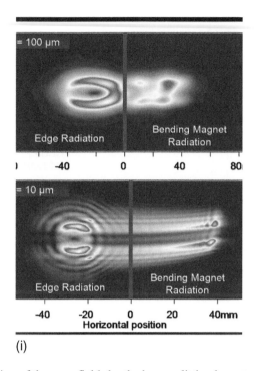

(i)

Fig. 11. (i) Partitioning of the wavefields by the beamsplitting box at wavelengths of 10 and 100 μm. The optimum place for this partitioning is 18 000 mm from the diamond window.

Mechanical vibrations in the mirror systems, particularly M1, lead to a degradation of performance, and care has been taken to minimize the effect of these vibrations. An analysis of the effects of vibrations from external sources on the mirror M1 has been discussed elsewhere (Creagh *et al.*, 2005). The mirror M1 for horizontal reflection of the infrared is a rectangular bar of $30 \times 54 \times 460$ mm, with the reflective face approximately 85 mm long at an angle of 55° to the vacuum chamber straight. A 3-mm slot in the mirror allows the high-energy radiation to pass through, as shown below. The first two resonant modes, calculated using finite element analysis (ANSYS Workbench 8.0), are at 181 (pure vertical motion) and 212 Hz (pure horizontal motion). Measurements have been made of the floor vibrations as they exist at present. They are ± 0.05 and ± 0.01 μm for the first and second harmonics, respectively. The motion of the mirror in the fundamental mode is in the vertical plane, and will not influence the performance of the mirror. The second harmonic vibration is in the plane of the mirror, and has the capacity to influence the beam optics because of the effect this has on the phase of the reflected wavefronts. (The wavelength selection in the instruments is made using Michelson interferometers. Motion in the horizontal plane can generate phase noise in the overall system.) Other sources of vibration are associated with the water cooling of the storage ring in general, and the dipole chamber in particular. Care has been taken to decouple the mirror beam from these sources of radiation. Special care has been taken to provide damping of the support for the mirror beam.

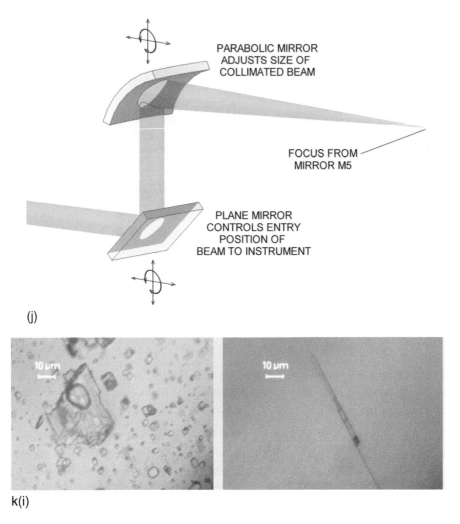

Fig. 11. (j) Schematic diagram of a matching optics box. Its role is to match the input characteristics of the instrument to that of the incident beam. (k) (i) Inclusions in the gemstone rhodonite. The compositions of the liquid and gaseous phases were found to be: in the left image, 15% (methane and nitrogen gas), 80% (saline solution), and 5% (solid ilmenite); in the right image, 30% (methane and nitrogen gas), 50% (saline solution), and 20% (solid ilmenite).

Calculations have been made of the heat load on the mirror. If the mirror were solid, it would intercept all the photon flux, including the X-ray component. This would lead to very significant heating. The heat load on the mirror is very dependent on the size of the slot and the accuracy of the positioning of the mirror relative to the electron beam. Distortion of the mirror due to heating would have a deleterious effect on the quality of the

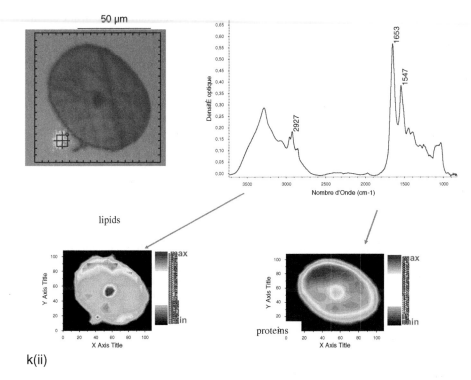

Fig. 11. (k) (ii) A diagram adapted from a PowerPoint slide (Cotte *et al.*, 2005) that shows distribution maps of lipid and protein in a cross section of the hair of a Turkmenistan mummy.

IR beam. Normally, the mirror M1 would be retracted during the initial beam injection cycle. However, if the storage ring is to run in the "top-up" mode, electron beam miss-steer values of ±1.0 mm are possible. Although data will not be taken during "top-up", the heat load on the mirror is 85 W, taken almost entirely on one half of the mirror. The mirror cooling system must be able to cater to heat loads of this magnitude. It should be noted, however, that in circumstances of beam instability of this kind, experiments would not proceed at the IR experimental stations. The slot in the mirror should also be slightly wider at the rear of the mirror, so that the front edge of the slot is the limiting aperture for the incident radiation. Consideration is being given to the determination of the response time of the system and whether the cooling water could be regulated by a temperature sensor or sensors attached in close proximity to the mirror face. In cases of extreme miss-steer of the electron beam, the mirror M1 will be retracted from the vacuum vessel. Figure 11(e)(i) shows the mirror M1 in its inserted position. In its retracted position, the mirror is completely withdrawn from the dipole vacuum chamber, and the gate valve between the mirror housing and the dipole vacuum chamber is closed.

The heat rise calculation indicates that, in normal operation, the thermal loading is small. Although the heat load on the mirror is relatively low, water cooling may still be needed to minimize the thermal distortion of the mirror. A concentric tube heat exchanger similar to that of the dipole absorbers used in the storage ring can provide sufficient cooling capacity (Fig. 11(e)(ii)). The water cooling system will be totally independent of other water systems associated with the synchrotron to avoid water-borne transmission of vibration. Because water pumps can cause vibration, passive cooling systems such as thermosyphon are being considered.

5.5.2. Edge and bending-magnet radiation

Calculations for the flux from a bending magnet usually concentrate on the radiation emitted from the electron bunch as it is accelerated by the magnetic field of the bending magnet. Another source of radiation exists, however. This corresponds to radiation emitted as the electron bunch enters the fringing field of the bending magnet; this radiation is referred to as "edge radiation". Figure 11(f) illustrates the formation of the origins of bending magnet and edge radiation.

Because the edge radiation originates at the entry point to the magnetic field of the dipole magnets, and the bending-magnet radiation occurs within the central region of the bending magnet, two different optical sources with different emissivities as a function of wavelength exist. And, they are spatially separated. This means that, since the focussing element, the ellipsoidal mirror M2, has fixed characteristics (R_t = 2969 mm; R_s = 1485 mm), the images are formed at different locations with respect to the CVD window. The geometry, however, is such that both beams come to a focus close to the CVD window.

Figure 11(g) shows wavefront calculations for 3 μm radiation. At the CVD window, the image size is contained within a square of dimensions 4 mm × 4 mm. The calculations have been made using the SRW Code (Chubar, 2005).

Because the intention is to use the edge radiation for high-resolution spectroscopy and the bending-magnet radiation for microspectroscopy, optical elements having different characteristics will be required for each of these beamlines.

5.5.3. The overall configuration of an IR beamline

Figure 11(h)(i) shows the overall configuration of an IR beamline. Comprehensive details of this have been given elsewhere by Creagh et al. (2006b). The principal role of any photon delivery system is to deliver as many photons as possible to the aperture of the measuring equipment. In the case of IR beamlines at the ESRF, CLRC Daresbury, the Australian Synchrotron, etc., these photons have to be delivered to commercially available microscope–spectrometers (usually Bruker or Nicolet) and high-resolution spectrometers (usually Bruker).

Considering Fig. 11(h)(ii): on emerging from the diamond, the IR beam moves into a high-vacuum pipe and into a beamsplitting module. This module performs the function of separating the edge from the bending-magnet components and directing the separated beams either to the microscope or to the high-resolution spectrometer. Examples of the partitioning of the edge radiation from the bending-magnet radiation are shown in Fig. 11(i) for wavelengths of 10 and 100 μm. At 1800 mm from the CVD window, it is possible to divide the beams in such a manner as to divert approximately the same photon flux into each of the instruments.

The edge component is diverted left into the high-resolution spectrometer. The bending-magnet radiation is diverted right into the microscope–spectrometer system.

The IR beamline cabin has separate rooms for the high-resolution spectrometer, the IR microscope, and for the custom equipment that would be brought by experimenters to the "bare port" to undertake surface science experiments such as reflectivity and ellipsometry, and a preparation room containing microscopes, micromanipulators, and perhaps a Raman microscope, for the alignment and testing of equipment prior to mounting on the IR microscope. A separate preparation room for the handling of biological specimens is provided elsewhere in the Australian Synchrotron building.

As mentioned earlier, it is important that all the photons in the beam be able to be accepted by the mirrors within the microscope, and that the beam is focussed by the microscope onto the specimen in a true confocal manner. This implies that the beam that enters the interferometer is parallel, and the diameter of the beam matches the aperture of the microscope. The diameter of this aperture will change according to the numerical aperture used in the microscope. Both the instruments have their incident beam characteristics matched to the characteristics of the incident beam by a matching optics box.

To ensure that this matching can be met, a parabolic mirror, followed by a plane mirror, are placed in the beam at a distance from the focal point of mirror 5A such that the image size matches the size dictated by the numerical aperture of the microscope (Fig. 11(j)). A number of these parabolic mirrors will be required for each distance from the focal point of mirror 5A. These mirrors will be mounted in a vacuum enclosure that will be isolated from the incident beamline and the microscope when changes need to be made. Positioning and steering of the mirrors will be effected by remote control, using standard optical modules fitted with motor encoders. The parabolic mirrors will be fitted with clamping magnets and will be able to be rapidly and reproducibly interchanged.

This system is in use at the ESRF by Dumas (2005) and is planned for Synchrotron SOLEIL (http://www.esrf.fr; http://www.synchrotron-soleil.fr).

Inevitably, there will be some effect on the IR beam from both mechanical vibrations and instabilities in the orbit of the circulating electrons. It is necessary, therefore, to incorporate with both instruments a feedback system to nullify their effects. Designs have been developed for a system to minimize vibrations in the mirrors and variations in the plane of the orbit of the electrons in the storage ring. These will be based on the designs of Martin *et al.* (1998). A similar system has recently been installed at CLRC, Daresbury (http://www.srs.ac.uk/srs/).

It must be stressed here that the instruments used in IR spectromicroscopy and high-resolution spectroscopy are slightly modified versions of standard laboratory units such as the Bruker IFS 66 of IFS125HR. In the case of spectromicroscopy, the high brightness of the synchrotron makes possible the investigation of systems in which the spatial distribution of molecules is of interest. Also, a focal plane array system would assist in studying the time evolution of systems.

For details of the equipment, their specifications, and the detectors that may be used, please refer to:

- http://www.thermo.com/BURedirect/welcomeMsg/1,5107,26,00.html
- http://www.bruker.de/

5.5.4. Use of IR microscopy in cultural heritage studies

As Bertrand has mentioned in Chapter 2 of this volume, IR techniques have not been as widely used in cultural heritage studies as might be expected. This is due, in part, to the fact that reliable IR beamlines are only now coming into service. In principle, synchrotron radiation beamlines will do everything an experimentalist would like to do in the laboratory, and more, and do it faster. Because of the high brightness of synchrotron radiation sources, it is possible to study the processes taking place *in aquo*, experiments are being undertaken by cell biologists on living cells, investigating, for example, the effect of anti-cancer drugs on living cancer cells.

Salvadó *et al.* (2005b) have recently written a review article titled the "Advantages of the use of SR-FTIR microspectroscopy: applications to cultural heritage". On the websites of all the major synchrotron radiation sources, extensive documentation is provided on how their IR facilities have been used to perform a wide range of experiments.

I list here some experiments that have been undertaken in the past 3 years.

Milsteed *et al.* (2005) studied geological thin sections of the gemstone rhodonite. Microspectroscopy showed that the inclusions included gaseous, liquid, and solid phases. The compositions of the liquid and gaseous phases were found to be: in the left image, 15% (methane and nitrogen gas), 80% (saline solution) and 5% (solid ilmenite) in the right image, 30% (methane and nitrogen gas), 50% (saline solution) and 20% (solid ilmenite).

Salvado *et al.* (2002) used FTIR spectroscopy to identify copper-based pigments in James Huget's Gothic altar pieces. In a more recent work, Salvadó *et al.* (2005a) studied the nature of medieval synthetic pigments: the capabilities of single-reflectance infrared spectroscopy.

Cotte *et al.* (2005) have studied the skin of an Egyptian mummy by infrared microscopy, investigating the lipid and protein content. Shown in Fig. 11(k) is a diagram adapted from a slide that shows distribution maps of lipids and proteins in a cross section of the hair of a Turkmenistan mummy.

The work by Zhang *et al.* (2001) on the lifetime of lithium ion batteries gives hope that *in situ* spectroelectrochemical studies of the degradation of protective layers on metal substrates will be able to be undertaken in the near future.

5.6. X-ray tomography

In the first volume of this book series, Casali (2006) has authored a chapter on X-ray and neutron computed tomography. It is not my intention to duplicate the material in his chapter. Rather, I wish to introduce two new techniques being developed in the field of medical imaging that may have uses for the study of small objects of cultural heritage significance.

Conventional tomography forms its image through the absorption of a photon beam as it passes through a material. It measures in essence the decrease in amplitude of the elec-tromagnetic wavefield. But a wave is characterized by an amplitude and a phase. Until recently, the phase changes occurring in the transport of the photon beam through a mate-rial medium have not been used in the formation of tomographic images.

Both the techniques outlined below have applications in cultural heritage studies.

5.6.1. Spherical wave projection

This method (Gao *et al.*, 1998) uses the fact that synchrotron source sizes are small, and the wavefront is quasi-coherent. Figure 12(a)(i) shows a geometrical optical representation of the imaging of an absorptive object using an incoherent, extended source. Note the blurring of the image due to the separation of the object and the image. In what follows, the argument is not wavelength specific: the argument applies to polychromatic sources. Figures 12(a)(ii) and (iii) show schematically the phase and amplitude distortions that occur in the polychromatic spherical wave.

A small (10 μm diameter) polychromatic source (industrial fine focus tubes or a circular aperture on a white beam synchrotron radiation beamline) can provide a high level of

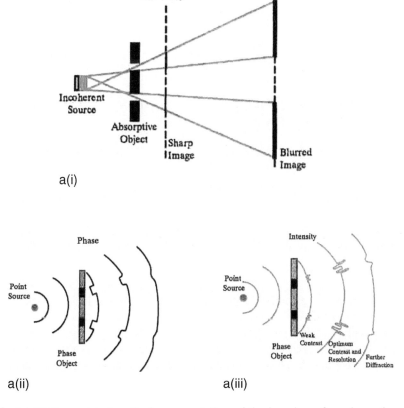

Fig. 12. (a) (i) Geometrical optical representation of the imaging of an absorptive object using an incoherent, extended source. Note the blurring of the image due to the separation of the object and the image. In what follows, the argument is not wavelength specific: the argument applies to polychromatic sources. (ii) Shows schematically the phase distortion that occurs in the polychromatic spherical wave. (iii) Shows schematically the amplitude distortion that occurs in the polychromatic spherical wave.

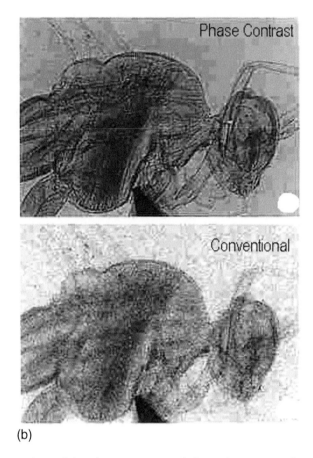

(b)

Fig. 12. (b) Comparison of the phase contrast and absorption contrast images of the head of a small dragonfly. Image provided with the permission of Dr. Stephen Wilkins.

spatial coherence: this is essential for this technique to work. The image is one of a class of in-line holograms that must undergo reconstruction to retrieve the spatial information (Pogany and Wilkins, 1997).

Figure 12(b) shows phase contrast and absorption contrast images of the head of a small dragonfly. It is clear that the phase contrast image shows more and much finer details than the conventional X-ray image. These images were taken with a laboratory fine-focus source. Of course, in this case, the object is not highly absorbing. But the integrity of the phase component of the image is only slightly compromised by the loss in intensity due to absorption by the sample, whereas the conventional technique becomes unusable when the absorption contrast becomes comparable to the minimum response function of the detector. A hologram of an object can be reconstructed successfully if the amplitudes are set to unity, and only the phases are used. In contrast, without the phase information, the hologram cannot be reconstructed successfully.

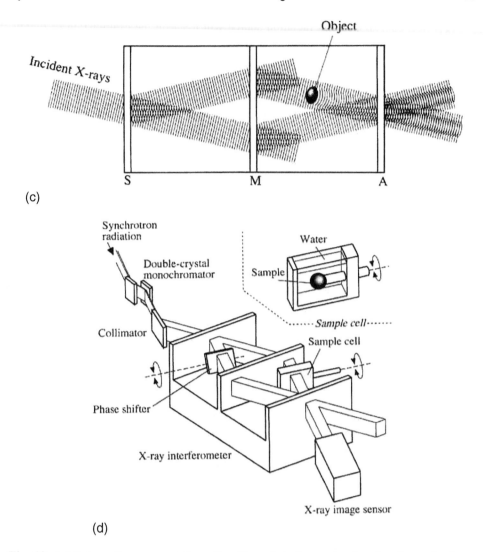

(c)

(d)

Fig. 12. (c) Schematic representation of an X-ray interferometer showing its use in phase measurements. S, M, and A are silicon wafers of equal thickness milled into a boule of [111] silicon. After milling the atomic planes in these wafers, maintain the same orientation as they had in the crystal boule. The wafers are cut such that the [110] direction lies parallel to the surface (in the plane of the diagram). The interferometer is then set to the angle of reflection of the [110] planes. If the thickness of the wafer is chosen correctly, the intensities of the transmitted and diffracted beams from S are equal. Similar behaviour occurs at M and A. Without an object in the beam, the transmitted and the diffracted beams have the same uniform intensity because the optical path lengths are equal. Placing an object in the beam causes the formation of a phase contrast image. (d) Experimental arrangement for phase-contrast X-ray-computed tomography.

Use of a bright synchrotron radiation source will assist in the examination of highly absorbing materials, such as microfossils embedded in a rock.

5.6.2. Interferometer techniques

The precise meaning of the term "phase contrast" differs with the method used. In an interferometer (Fig. 12(c)), fringes are seen for every 2π change in phase shift, whereas, in the holographic method, the fringes are due to Fresnel diffraction.

An X-ray interferometer can be used to give phase-shifted images on reasonably small objects. The limitation to resolution is the resolution of the detector and imperfections in the beamsplitting wafers. The method actually measures the integrated refractive index across the sample. See Section 3.1.1.

$$n = 1 - \delta - i\beta$$

δ is related to the real part of the X-ray scattering amplitude, and β to the linear attenuation coefficient μ_l. The refractive index of an object placed in one of the beam paths in the interferometer causes phase shifts to occur in the image formed in the exit beams. The width of the interferometer beams determines the lateral extent of the object that can be investigated. In the absence of the object, no contrast is observed in these beams. Creagh and Hart (1970) used an X-ray interferometer to measure the real part of the forward-scattering amplitude in lithium fluoride.

Momose et al. (1998) have reported a new system for phase contrast tomography (Fig. 12(d)). In this technique, a phase-shifting plate is inserted into one part of the beam path to act as a phase shifter, and the sample is placed in the beam path. Both can be rotated about an axis in the horizontal plane. Their method involves rotating the phase shifter and the sample in synchronism in steps of $2\pi/M$, where M is the number of interference patterns chosen. The phase-mapped image is then calculated from these M interference patterns. Note that this procedure removes the absorption contrast. Momose et al. (1998) used this technique to study sections of human tissue (sometimes containing cancerous material) 1 mm in diameter.

ACKNOWLEDGEMENTS

Throughout my career as a research scientist, I have enjoyed the love and support of my wife Helen. Without her encouragement and willingness to accept the foibles of a working scientist, nothing could have been accomplished.

I thank the research associates in my Cultural Heritage Research group for their excellent work. These include Dr.Vincent Otieno-Alego, Alana Lee, Maria Kubik, Drs. Denise and Peter Mahon, David Hallam, David Thurrowgood, Dr. Paul O'Neill, Dr. Stephen Holt, Dr. Ian Jamie, and Dr. Tim Senden. They have been a pleasure to work with. Others such as Metta Sterns have freely given their time to assist with the processing of results.

My synchrotron radiation research has been supported by the Australian Synchrotron Research Program. I would especially like to thank Drs. Garry Foran, James Hester, and

David Cookson for their assistance with the creation of the Australian National Beamline Facility and in the performance of experiments associated with my research projects.

Much of my research has been undertaken with funding from research bodies such as the Australian Research Council, the Australian Institute of Nuclear Science and Engineering, and with the collaboration of the National collecting institutions (the National Archives of Australia, the National Museum of Australia, the Australian War Memorial, the National Film and Sound Archive, the national Library of Australia, and the National Gallery of Australia).

Funding from the University of Canberra has assisted me in the creation of a first-rate spectroelectrochemical laboratory for cultural heritage and forensic science research.

The State Government of Victoria, through its sponsorship of the construction of the Australian Synchrotron, has given me opportunity to collaborate in a number of projects, in particular the project involving the design of the IR beamline. I would like to thank Prof. John Boldeman, Andy Broadbent, Jonathon McKinlay, and the beamline scientists Drs. Mark Tobin, Chris Glover, and Kia Wallwork for their friendship and for their assistance in providing material for this manuscript. I am very grateful to Drs. Paul Dumas (Synchrotron Soleil) and Michael Martin (Advanced Light Source) for their friendship and the assistance that they have very generously given me throughout the IR beamline project.

REFERENCES

Amemiya, Y., Arndt, U.W., Buras, B., Chikawa, J., Gerward, L., Langford, J.L., Parrish, W., De Wolf, P.M., 2004. Detectors for X-rays. In *International Tables for Crystallography*, Volume C, Edition 4, Section 7. Ed. Prince, E. International Union of Crystallography, Kluwer: Dordrecht, pp. 612–632.

ANSYS Workbench 8.0, Finite Element Analysis Software: http://www.ansys.com

Atwood, D., 1999. *Soft X-rays and Extreme Ultraviolet Radiation*. Cambridge University Press: Cambridge, UK.

Balaic, D.X., Nugent, K.A., 1995. The X-ray optics of tapered capillaries. *Appl. Opt.* **34**, 7263–7272.

Balaic, D.X., Nugent, K.A., Barnea, Z., Garrett, R.F., Wilkins, S.W., 1995. Focusing of X-rays by total external reflection from a paraboloidally tapered glass capillary. *J. Synchrotron Radiat.* **2**, 296–299.

Balaic, D.X., Barnea, Z., Nugent, K.A., Varghese, J.N., Wilkins, S.W., 1996. Protein crystal diffraction patterns using a capillary-focused synchrotron X-ray beam. *J. Synchrotron Radiat.* **3**, in press.

Beaumont, J.H., Hart, M., 1974. Multiple-Bragg reflection monochromators for synchrotron radiation. *J. Phys. E: Sci. Instrum.* **7**, 823–829.

Berman, L.E., Hart, M., 1991. Adaptive crystal optics for high power synchrotron sources. *Nucl. Instrum. Methods* **A302**, 558–562.

Bilderback, D.H., Theil, D.J., Pahl, R., Brister, K.E., 1994. X-ray applications with glass-capillary devices. *J. Synchrotron Radiat.* **1**, 37–42.

Bonse, U., Hart, M., 1965. Tailless X-ray single-crystal reflection curves obtained by multiple reflection. *Appl. Phys. Lett.* **7**(9), 238–240.

Bornebusch, H., Clausen, B.S., Stefessen, G., Lützenkirchen-Hecht, D., Frahm, R., 1999. A new approach for QEXAFS data acquisition. *J. Synchrotron Radiat.* **6**, 209–211.

Buras, B., David, W.I.F., Gerward, L., Jorgensen, J.D., Willis, B.T.M., 1994. Energy dispersive techniques. In *International Tables for Crystallography*, Volume C, Edition 3, Section 2.5.1. Ed. Prince, E. International Union of Crystallography, Kluwer: Dordrecht, pp. 84–87.

Casali, F., 2006. X-ray and neutron radiography and computed tomography for cultural heritage. In *Physical Principles in Art and Archaeology*. Eds. Bradley, D.A., Creagh, D.C. Elsevier: Amsterdam, The Netherlands.

Chantler, C., 2002–2006. http://optics.ph.unimelb.edu.au/~chantler/home.html

Chavanne, J., 2002. Review of undulators at the ESRF. www.aps.anl.gov/News/Conferences/ 2002/novel_ids/Joel_Chavanne.pdf

Chubar, O., 2005. Simulation of emission of the propagation of synchrotron wavefronts using the methods of physical optics. In *The Proceedings of the International Conference on Infrared Microscopy and Spectroscopy using Accelerator Based Sources*, WIRMS, Rathen, Germany, 26–30 June 2005.

Cotte, M., Walter, P., Tsoucaris, G., Dumas, P., 2005. Studying skin of an Egyptian mummy by infrared microscopy. *Vib. Spectrosc.* **38**(1–2), 159–167.

Creagh, D.C., 2004a. X-ray absorption spectra. In *International Tables for Crystallography*, Volume C, Edition 4, Section 4.2.3. Ed. Prince, E. International Union of Crystallography, Kluwer: Dordrecht, pp. 212–220.

Creagh, D.C., 2004b. X-ray absorption (or attenuation) coefficients. In *International Tables for Crystallography*, Volume C, Edition 4, Section 4.2.4. Ed. Prince, E. International Union of Crystallography, Kluwer: Dordrecht, pp. 220–236.

Creagh, D.C., 2004c. Filters and monochromators. In *International Tables for Crystallography*, Volume C, Edition 4, Section 4.2.5. Ed. Prince, E. International Union Crystallography, Kluwer: Dordrecht, pp. 236–241.

Creagh, D.C., 2004d. X-ray dispersion corrections. In *International Tables for Crystallography*, Volume C, Edition 4, Section 4.2.6. Ed. Prince, E. International Union of Crystallography, Kluwer: Dordrecht, pp. 242–258.

Creagh, D.C., 2005. The use of radiation for the study of artifacts of cultural heritage significance. In *Physics for the Preservation of Cultural Heritage*. Eds. Fernandez, J.E., Maino, G., Tartari, A. CLUEB: Bologna, pp. 35–67.

Creagh, D.C., Ashton, J.A., 1998. The use of X-ray analysis techniques for the study of materials of cultural heritage significance in museums and galleries. In *Proceedings of the European Conference on Energy Dispersive X-ray Spectrometry*. Eds. Fernandez, J.E., Tatari, A. Editrice Compositori: Bologna, Italy, pp. 299–303.

Creagh, D.C., Hart, M., 1970. X-ray interferometric measurements of the forward scattering amplitude in lithium fluoride. *Phys. Stat. Solid.* **37**, 753–758.

Creagh, D.C., Garrett, R.F., 1995. Testing of a sagittal focussing monochromator at BL 20B at the photon factory. In *Access to Major Facilities Program*. Ed. Boldeman, J.W. ANSTO: Sydney, Australia, pp. 251–252.

Creagh, D.C., Martinez-Carrera, S., 2004. Precautions against radiation injury. In *International Tables for Crystallography*, Volume C, Edition 4, Chapter 10. Ed. Prince, E. International Union of Crystallography, Kluwer: Dordrecht, pp. 949–959.

Creagh, D.C., Otieno-Alego, V., 2003. The use of radiation for the study of materials of cultural heritage significance. *Nucl. Instrum. Methods Phys. Res. B* **213C**, 670–676.

Creagh, D.C., Otieno-Alego, V., O'Neill, P.M., 1999. X-ray reflectivity studies and grazing incidence X-ray diffraction studies of the adhesion of protective wax coatings on metal surfaces. *Reports, Australian Synchrotron Research Program*. ANSTO: Sydney, Australia.

Creagh, D.C., McKinlay, J., Dumas, P., 2005. The design of the IR beamline at the Australian synchrotron. *Radiat. Phys. Chem.* Accepted for publication.

Creagh, D.C., Kubik, M.E., Syerns, M., 2006a. A feasibility study to establish the provenance of Australian aboriginal artifacts using synchrotron radiation X-ray diffraction and PIXE. *Nucl. Instrum. Methods Phys. Res. A* To be published.

Creagh, D.C., McKinlay, J., Tobin, M., 2006b. The current status of the IR beamline at the Australian synchrotron. In *Synchrotron Radiation Instrumentation Conference 2006*, Daegu, Korea.

Creagh, D.C., Thorogood, G., James, M., Hallam, D.L., 2004. Diffraction and fluorescence studies of bushranger armour. *Rad. Phys. Chem.* **71,** 839–840.

Creagh, D.C., Foran, G.J., Cookson, D.J., Garrett, R.F., Johnson, F., 1998. An eight-position capillary sample spinning stage for the diffractometer at beamline 20B at the Photon Factory. J. Synchrotron Rad. **5**, 823–825.

De Marco, R., 2003. *In situ* X-ray diffraction studies of the sensor/electrolyte interface in environmental samples. http://synchrotron.vic.gov.au/files/documents/Roland_De_Marco_Presentation_Feb_2003.pdf

De Ryck, I., Adriens, A., Pantos, E., Adams, F., 2003. A comparison of microbeam techniques for the analysis of corroded ancient bronze objects. *Analyst* **128**, 1104–1109.

Duke, P., 2000. *Synchrotron Radiation*. Oxford University Press: Oxford, UK.

Dumas, P., 2005. Synchrotron infrared emission and spectroscopic applications. In *Lectures to Hercules Workshop*, ESRF, Grenoble, France, 20 February–25 May 2005.

Engstrom, P., Rindby, A., Vincze, L., 1996. Capillary optics. *ESRF Newsletter* **26**, 30–31.

Foran, G.J., 2004. Ionization states in iron oxides. Private communication.

Freund, A.K., 1993, Thin is beautiful, *ESRF Newsletter* **19**, 11–13.

Gao, D, Gureyev, T.E, Pogany, A., Stevenson, A.W., Wilkins, S.W., 1998. New methods of X-ray imaging based on phase contrast. In *Medical Applications of Synchrotron Radiation*. Eds. Ando, M., Uyama, C. Springer: Tokyo, Japan, pp. 63–71.

Giles, C., Vettier, C., De Bergevin, F., Grubel, G., Goulon, J., Grossi, F. 1994. X-ray polarimetry with phase plates. *ESRF Newsletter* **21**, 16–17.

Glatter, O., May, R., 2004. Small-angle techniques. In *International Tables for Crystallography*, Volume C, Edition 3, Section 4.2.4. International Union of Crystallography, Kluwer: Dordrecht, pp. 89–112.

Glover, C.J. McKinlay, J., Clift, M., Barg, B., Broadbent, A., Boldeman, J.W., Ridgway, M., Foran, G.J., Garrett, R.F., Lay, P., 2006. The X-ray absorption spectroscopy beamline at the Australian synchrotron. In *Synchrotron Radiation Instrumentation Conference 2006*, Daegu, Korea.

Guerra, M.F., Calligaro, M., Radtke, M.N., Reiche, I., Riesemeier, H., 2005. Fingerprinting ancient gold by measuring Pt with spatially resolved high energy Sy-XRF. *Nucl. Instrum. Methods B* **240**, 505–511.

Hallam, D.L., Adams, C.D., Creagh, D.C., Holt, S.A., Wanless, E.J., Senden, T.J., Heath, G.A., 1997. Petroleum sulphonate corrosion inhibitors. In *Metal 1995*. Eds. McLeod, I.D., Pennec, S.L., Robbiola, L. James and James Scientific Publishers: London, UK.

Hanfland, M., Haussermann, D., Snigirev, A., Snigireva, I., Ahahama, Y., McMahon, M., 1994. Bragg -Fresnel lens for high pressure studies. *ESRF Newsletter* **22**, 8–9.

Hannon, F.E., Clarke, J.A., Hill, C, Scott, D.J., Shepherd, B.J.A., 2004. Construction of an Apple II type undulator at the Daresbury Laboratory for the SRS. In *Proceedings of EPAC*, Lucerne, Switzerland.

Hart, M., 1971. Bragg reflection X-ray optics. *Rep. Prog. Phys.* **34**, 435–490.

Hart, M., Rodriguez, A.R.D., 1978. Harmonic-free single-crystal monochromators for neutrons and X-rays. *J. Appl. Cryst.* **11**(4), 248–253.

Hashizume, H., 1983. Asymmetrically grooved monolithic crystal monochromators for suppression of harmonics in synchrotron X-radiation. *J. Appl. Cryst.* **16**, 420–427.

Hoffmann, A., 2004. *Synchrotron Radiation*. Cambridge University Press: Cambridge, UK.

Holt, S.A., Brown, A.S., Creagh, D.C., Leon, R., 1996. The application of grazing incidence X-ray diffraction and specular reflectivity to the structural investigation of multiple quantum well and quantum dot semiconductor devices. *J. Synchrotron Radiat.* **4**, 169–174.

Jennings, L.D., 1981. Extinction, polarization and crystal monochromators. *Acta Cryst.* **A37**, 584–593.

Kennedy, C.J., Wess, T.J., 2006. The use of X-ray scattering to analyze parchment structure and degradation. In *Physical Techniques for the Study of Art, Archaeometry and Cultural Heriatage*. Eds. Bradley D.A., Creagh, D.C. Elsevier: Amsterdam, The Netherlands.

Kikuta, S., 1971. X-ray crystal monochromators using successive asymmetric diffractions and their applications to measurements of diffraction curves. II. Type 1 collimator. *J. Phys. Soc. Jap.* **30**(1), 222–227.

Kikuta, S., Kohra, K., 1970. X-ray crystal collimators using successive asymmetric diffractions and their applications to measurements of diffraction curves. I. General considerations on collimators. *J. Phys. Soc. Jap.* **29**(5), 1322–1328.

Kohsu Seiki Company, 2006. http://www.kagaku.com/kohzu/english.html

Krause, M.O., 1979. Tale of X-ray fluorescence yields. *J. Phys. Chem. Ref. Data* **8**, 307–399.

Kumakhov, M.A., 1990. Channeling of photons and new X-ray optics. *Nucl. Instrum. Meth. Phys. Res.* B **48**, 283–286

Kumakhov, M.A., Komarov, F.F., 1990. Multiple reflection from surface X-ray optics. *Phys. Rep.* **191**, 289–350.

Lee, A.S., Mahon, P.J., Creagh, D.C., 2006. Raman analysis of iron gall inks on parchment. *J. Vib. Spectrosc.* **41**, 170–175.

Leyssens, K., Adriaens, A., Dowsett, M.G., Schotte, B., Oloff, I., Pantos, E., Bell, A.M.T., Thompson, M., 2005. Simultaneous in situ time resolved SR-XRD and corrosion potential analyses to monitor the corrosion on copper. *Electrochem. Comm.* **7**(12), 1265–1270.

MAR Detector Systems, 2006. http://www.mar-usa.com/products/marccd.htm

Martin, M.C., Byrd, J.M., McKinney, W.R. 1998. High frequency electron beam motion observed and resolved. In *Advanced Light Source*. Lawrence Berkeley National Laboratory, Contract DE-AC03-76P00098.

Martinetto, P., Anne, M., Dooryhee, E., Tsoucaris, G., Walter, P, 2000. A synchrotron X-ray diffraction study of Egyptian cosmetics. In *Radiation in Art and Archaeometry*. Eds. Creagh, D.C., Bradley, D.A. Elsevier: Amsterdam, The Netherlands.

Matsushita, T., Kikuta, S., Kohra, K., 1971. X-ray crystal monochromators using successive asymmetric diffractions and their applications to measurements of diffraction curves. III. Type II collimators. *J. Phys. Soc. Jap.* **30**(4), 1136–1144.

Matsushita, T., Ishikawa, T., Oyanagi, H., 1986. Sagitally focusing double-crystal monochromator with constant exit height at the photon factory. *Nucl. Instrum. Methods* **A246**, 377–379.

Milsteed, P.W., Mermagh, T.P, Otieno-Alego, V., Creagh, D.C., 2005. Inclusions in the transparent gem rhodonite from Broken Hill, New South Wales, Australia. *Gems Gemology*, **41**(3), 247–254.

Momose, A., Takeda, T., Yoneyama, A., Hirano, K., 1998, Perspectives for medical applications of phase-contrast X-ray imaging. In *Medical Applications of Synchrotron Radiation*. Eds. Ando, M., Uyama, C. Springer: Tokyo, Japan, pp. 54–62.

Muller, M., Czhak, C., Burghamme, M., Riekel, C., 2000. The small angle scattering and fibre diffraction produced on single cellulose fibres. *J. Appl. Crystallogr.* **33**, 235–248.

Muller, M., Papiz, M.Z., Clarke, D.T., Roberts, M.A., Murphy, B.M., Burghammer, M., Riekel, C., Pantos, M., Gummeweg, J., 2004. Identification of ancient textile fibres from Khirbet Qumran caves using synchrotron radiation microbeam diffraction. *Spectrochim. Acta B* **59**, 1669–1674.

O'Neill, P.M., Creagh, D.C., Sterns, M., 2004. Studies of the composition of pigments used traditionally in Australian aboriginal bark paintings. *Radiat. Phys. Chem.* **71**(3–4), 841–842.

Oshima, K., Harada, J., Sakabe, N., 1986. Curved crystal monochromator. In *X-ray Instrumentation for the Photon Factory: Dynamic Analyses of Micro-Structures in Matter*. Eds. Hosoya, S., Iitaka Y., Hashizume, H. pp. 35–41.

OSMIC^{TM}, 1996. Catalogue. In *A New Family of Collimating and Focusing Optics for X-ray Analysis*. Michigan, USA.

Otieno-Alego, V., Heath, G.A., Viduka, A., Hallam, D.L., Creagh, D.C., 1999. Evaluation of the corrosion performance of petroleum sulphonates as additives in wax coatings. In *Metal 98*. Eds. Mourey, W.A., McLeod, I. James and James Scientific Publishers: London, pp. 159–161.

Owens, M., Ullrich, J., et al., 1996. Polycapillary X-ray optics for macromolecular crystallography. SPIE **2859**, 200–209.

Pantos, E., 2006. http://srs.dl.ac.uk/people/pantos/previous_apointments.html

Pantos, E., Bertrand, L., 2006. Compilation of papers in the fields of archaeology, archaeometry, and conservation science which use synchrotron radiation techniques. http://srsdl.ac.uk/arch/publications.html

Pantos, E., Tang, C.G., MacLean, E.J., Cheung, K.C., Strange, R.W., Rizkallah, P.J., Papiz, M.Z., Colston, S.L., Roberts, M.A., Murphy, B.M., Collins, S.P., Clark, D.T., Tobin, M.J., Zhilin, M., Prag, K., Prag, A.J.N., 2002. Applications of synchrotron radiation to archaeological ceramics. In *Modern Trends in Scientific Studies on Ancient Ceramics*, Volume 1011. Eds. Kilikoglou, V., Hein, A., Maniatis, Y. BAR International Series: pp. 377–384.

Paterson, D., Boldeman, J.W., Cohen, D., Ryan, C.G., 2006. Microspectroscopy beamline at the Australian synchrotron. In *Synchrotron Radiation Instrumentation Conference 2006*, Daegu, Korea.

Peele, A.G., Nugent, K.A., Rode, A.V., Gabel, L.K., Richardson, M.C.M., Strack, R., Siegmund, W., 1996. X-ray focusing with lobster-eye optics: a comparison of theory with experiment. *Appl. Opt.* **35**(22), 4420–4425.

Pogany, A., Wilkins, S.W., 1997. Contrast and resolution in imaging with a microfocus X-ray source. *Rev. Sci. Instrum.* **68**, 2774–2782.

QUANTUM, 2006. http://www.adsc-xray.com/Q4techspecs.html

Rietveld, H., 1967. A technique for whole pattern crystal refinement. *Acta Crysrallogr.* **22**, 151–152.

Salvadó, N., Pradell, T., Pantos, E., Papiz, M.Z., Molera, J., Seco, M., Ventrall-Saz, M., 2002. Identification of copper-based pigments in James Huget's Gothic altar pieces by Fourier Transform Infrared spectroscopy and synchrotron radiation X-ray diffraction. *J. Synchrotron Radiat.* **9**, 215–222.

Salvadó, N., Butì, S., Tobin, M.J., Pantos, E.J., Prag, N.W., Pradell, T., 2005a. The nature of medieval synthetic pigments: the capabilities of SR-infrared spectroscopy. In *IRUG 6 Proc.*, Firenze, 29 March–1 April 2004.

Salvadó, N., Butì, S., Tobin, M.J., Pantos, E.J., Prag, N.W., Pradell, T., 2005b. Advantages of the use of SR-FTIR microspectroscopy: applications to cultural heritage. *Anal. Chem.* **77**(11), 3444–3451.

Sasaki, S., 1994. Analyses for a planar variably-polarizing undulator. *Nucl. Instrum. Methods Phys. Res. A* **347**, 310–320.

Siemens, 1996a. Parallel beam optics for measurements of samples with irregularly shaped surfaces. *Lab Report X-ray Analysis, DXRD*, **13**, Siemens: Karlsruhe.

Siemens, 1996b. Goebel mirrors for X-ray reflectometry investigations. *Lab Report X-ray Analysis, DXRD*, **14**, Siemens: Karlsruhe.

Siemens, 1996c. Grazing incidence diffraction with Goebel mirrors. *Lab Report X-ray Analysis, DXRD*, **15**, Siemens: Karlsruhe.

Smith, M., Fankhauser, B., Kubik, M.E., Tait, R., 2007. Characterization of Australian red ochre deposits: I. a review of major sources, their geomorphology, and archaeology. *Archaeometry* In preparation.

Snigirev, A., 1996. Bragg–Fresnel optics: new fields of applications. *ESRF Newsletter* **22**, 20–21.

Stephens, P.W., Eng, P.J., Tse, T., 1992. Construction and performance of a bent crystal X-ray monochromator. *Rev. Sci. Instrum.* **64**(2), 374–378.

Sussini, J., Labergerie, D. 1995. Bi-morph piezo-electric mirror: a novel active mirror. *Synchrotron Radiat. News* **8**(6), 21–26.

Tobin, M., 2006. Private communication.

Warren, B.E., 1968. *X-ray Diffraction*. Addison-Wesley: New York.

Wess, T.J., Drakopoulos, M., Snigirev, A., Wouters, J., Paris, O., Fratzl, P., Hiller, J., Nielsen, K., 2001. The use of small angle X-ray diffraction studies for the analysis of structural features in archaeological samples. *Archaeometry* **43**, 117–129.

Williams, G.P., 1990. The initial scientific program at the NSLS. *Nucl. Instrum. Methods Phys. Res. A* **291**, 8–13.

XFIT, 2006. http://www.ansto.gov.au/natfac/asrp7_xfit.html

Young, R.A. 1963. Balanced filters for X-ray diffractometry. *Z. Kristallogr.* **118**, 233–247.

Young, R.A. 1993. *The Rietveld Method*. Oxford University Press: New York.

Zhang, X., Ross Jr., P.N., Kostecki, R., Kong, F., Sloop, S., Kerr, J.B., Striebel, K., Cairns, E.J., McLarnon, F., 2001. Diagnostic characterization of high power lithium-ion batteries for use in hybrid electric vehicles. *J. Electrochem. Soc.* **148**(5), A463–A470.

Chapter 2

Synchrotron Imaging for Archaeology, Art History, Conservation, and Palaeontology

L. Bertrand

Synchrotron SOLEIL, HALO – Heritage and Archaeology Liaison Office,
L'Orme des merisiers, Saint-Aubin BP48, F-91192, Gif-sur-Yvette cedex, France
Email: loic.bertrand@synchrotron-soleil.fr

Abstract

This chapter surveys the use of synchrotron radiation in archaeology, art history, conservation science, and palaentology. It follows on from the preceeding introduction on synchrotron radiation given in Chapter 1 (Professor Dudley Creagh). After giving a brief survey of the current use of synchrotron radiation the author describes both invasive and non-invasive techniques for the examination of artefacts and their environment. Examples are given in the field of conservation science through the study of the alteration of iron artefacts and waterlogged wood using micro-X-ray spatial fluorescence mapping, and in that of paleontology with the description of micro-tomographic studies of bones of a primate from the Late Eocene period are described. An introduction is given to the complementary techniques of synchrotron infrared Fourier Transform micro-spectroscopy, micro-X-ray absorption near edge structure, and micro-X-ray diffraction.

Keywords: Synchrotron, imaging, archaeology, art, conservation science, palaeontology, X-rays.

Contents

1. INTRODUCTION

Ancient materials, most often handmade from natural ingredients and sometimes heavily altered due to long-term ageing, are highly heterogeneous. Additionally, art materials are usually subject to treatments and modifications to play with their optical appearance, which is strongly influenced by the material texture at the micrometre length scale (*i.e.* wavelengths in the visible range). The ability to resolve chemical and structural information concerning these local heterogeneities (mineral grains, inclusions, surfaces and interfaces, cracks, and micro-structural defects) is then of particular relevance.

In addition, the chemistry of the systems under consideration is usually very complex. Organic and inorganic compounds, their mixtures, alteration and interaction products have to be understood. Complementary to topographic imaging and mere elemental identification, chemical selectivity is required to attain a satisfactory understanding of the materials. The three information levels have to be correlated on a sample to cope with its heterogeneity.

Most often, the relevant heterogeneity length scale falls into the 100 nm to 10 μm range (for an example, see Fig. 1). This *mesoscopic* length scale is intermediate between macroscopic (accessible through naked-eye observations, stereomicroscopy optical microscopy, etc.,

Fig. 1. (a) Photography of a fragment of an iron of the Second Iron Age undergoing catastrophic corrosion process. The object was originally cleaned for exhibition without further stabilisation treatment. Corrosion leads to the cracking and peeling of the corrosion layers, destroying precious archaeological information (shape and ornamentation) (Bertholon, 2001). (b) Optical microscopy image of a cross section from a corroded iron artefact. The distribution of the trace phases needs to be understood at a length scale smaller than 10 μm. Images are courtesy of R. Bertholon and S. Réguer, respectively.

which are widely used by the heritage community) and truly microscopic levels (electron microscopes, etc.). As presented earlier, synchrotron radiation techniques are particularly efficient to obtain chemically selective information at such length scales and to obtain 2D or 3D maps of elemental, chemical, and structural information.

2. CURRENT USE OF SYNCHROTRON IMAGING ON ANCIENT MATERIALS

Laboratories working on the analysis of ancient materials are heavy users of infrared spectroscopy, X-ray diffraction, X-ray fluorescence, and radiography, which are all among the top ten methods used by these institutions (see Fig. 2). These methods find direct counterparts at synchrotron sources with improved analytical sensitivity, ultra-short acquisition times, and a micrometre-range resolution. Indeed, even hard X-ray beams can now be focussed down to micrometric or sub-micrometric spots by using synchrotron sources.

In this chapter, imaging is understood in its broad meaning, including both raster scanning and full-field methods, leading to the collection of 2D and 3D maps with a micrometre-level resolution. In total, this represents about one-fifth of the total synchrotron publications for the heritage field known to us.[1] The main techniques dealt with are listed in Table 1, and the reader is invited to refer the chapter on Synchrotron Radiation

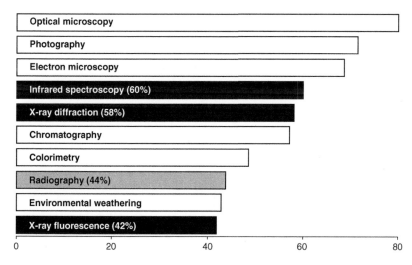

Fig. 2. Most widely used methods in European laboratories working on ancient materials. Methods that find a direct counterpart at synchrotron facilities are shown in black and grey. Adapted from the LabsTech 2003 data (LabsTech European Coordination Network (1999–2002)).

---

[1] Synchrotron SOLEIL Heritage and Archaeology Liaison Office – reference database website available at http://www.synchrotron-soleil.fr/heritage/

Table 1. List of synchrotron imaging techniques in scanning and full-field modes

Method	Corresponding information
(a) Scanning experiments	
Wide-angle X-ray diffraction and scattering (μXRD and μWAXS)	Crystalline structure
Small-angle X-ray scattering (μSAXS)	Organisation at the nanometre scale
X-ray fluorescence (μXRF)	Elemental content
X-ray absorption (μXAS, μXANES, and μEXAFS)	Oxidation state and local environment
Fourier-transform infrared spectromicroscopy (μFTIR)	Chemical bond identification
(b) Full-field experiments	
Micro-computed X-ray tomography (μCT)	3D radiography
X-ray microscopy (XRM)	Radiography

and its Use in Art, Archaeometry, and Cultural Heritage Studies (D. Creagh, Chapter 1) for an introduction on the corresponding methods.

A survey of the synchrotron heritage publications clearly shows that the use of synchrotron techniques in *microfocussed mode* exceeds that of large beam. Microfocussed beam usage ranges from 100% for FTIR to 49% for XRD (see Fig. 3). Microbeams are either used for single-spot acquisitions on minute samples or to acquire 2D (3D) raster scans. Such scans

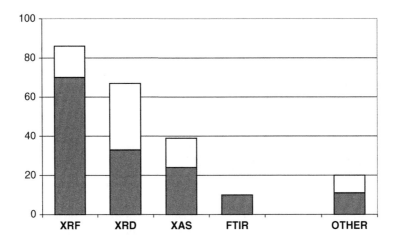

Fig. 3. Distribution of the synchrotron publications on heritage according to the technique used. Microbeam experiments are shown in dark grey. For experimental method abbreviations, please refer to Table 1 (data time span 1986–2005). A list of synchrotron publications on heritage is maintained at the SOLEIL synchrotron website (http://www.synchrotron-soleil.fr/heritage/).

can lead to very precise mapping of composition, structure, and chemical information at a micrometre-length-scale level, crucial for the understanding of the materials, their ageing, and the treatments applied to them. Using multitechnique synchrotron beamlines, elemental, chemical, and structural information can be collected at the same acquisition point. Reducing the beam footprint can additionally decrease the complexity due to the number of chemical species contributing to each spectrum collected, thus simplifying further data processing. This is a very important factor for chemical speciation of heterogeneous samples through X-ray absorption spectroscopy, and is used extensively for other fields such as environmental sciences (Isaure *et al.*, 2002; Manceau *et al.*, 2002). Comparison and clustering of individual spectra can then contribute to the understanding of the underlying compositional and structural correlations. However, reducing the spot size diminishes the representativeness of the experiment. It can also be at the origin of specific crystal orientation issues in 2D monochromatic μXRD mapping when the spot size reaches that of the crystallite. Very few, if any, crystal plans are in a good orientation to diffract, and almost no spot is observed in the diffraction image. In such a situation, polychromatic ("white") beam techniques are preferred.

Alternatively, *full-field* imaging methods rely on the interaction between a wide beam and the sample: X-ray microscopy, X-ray micro-computed tomography, etc. The resolution is then primarily limited by that of the detector and can be less than a micrometre at synchrotron tomography beamlines. Some of the major specificities of the synchrotron radiation (high monochromaticity, intensity, nearly parallel geometry, and coherence) are particularly suited to the observation of samples that are difficult to image using conventional radiography and microtomography equipment (Salvo *et al.*, 2003). In particular, the contrast of materials that have a rather homogeneous density distribution can be strongly enhanced using phase contrast imaging, and the *beam hardening* effect, due to the polychromaticity of laboratory X-ray sources, is suppressed. These two points are essential for palaeontological research (Tafforeau, 2004; Tafforeau *et al.*, 2006). Recent additional solutions to increase the contrast, developed primarily for biomedical synchrotron analysis, include diffraction-enhanced imaging (DEI).

The main advantages sought by current synchrotron users from the heritage community are therefore:

1. the use of microfocussed setups for methods that they also generally have access to using laboratory equipment; and
2. the access to methods that are specific to synchrotron radiation, as X-ray absorption that requires the tuneable and highly monochromatic beam delivered at synchrotron facilities.

Most heritage users rely on a conjunction of synchrotron and non-synchrotron imaging techniques. The latter can be elementally or chemically selective, such as Raman microscopy, scanning electron microscopy coupled to energy dispersive spectrometry (SEM-EDX), transmission electron microscopy (TEM), laboratory X-ray techniques (XRF, XRD), FTIR, ion beam analyses (IBA: PIXE, etc.), etc. In particular, Raman microscopy is currently leading to unprecedented developments for the understanding of art and archaeological samples at the micrometer-scale level (Neff *et al.*, 2004) and can now be coupled to synchrotron characterisations (Davies *et al.*, 2005). To date, cross-technique correlation remains difficult, and more research has to be done to attain a satisfactory complementary use of synchrotron and non-synchrotron techniques.

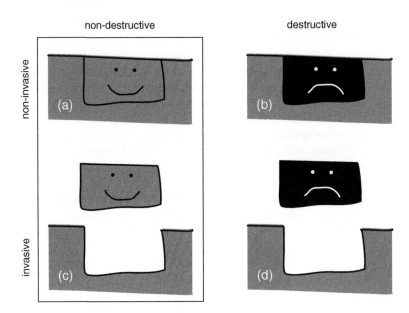

Fig. 4. Schematic distinction between invasive/non-invasive (sampling required) and destructive/non-destructive characterisations. Using appropriate precautions, synchrotron techniques fall in the non-destructive category.

3. EXPERIMENTAL REQUIREMENTS

Imaging an object or a sample in 2D or 3D implies specific experimental requirements to be met. On the one hand, if sampling is allowed, the experiment is usually performed invasively (with sampling, see Fig. 4), and almost all the experimental methods evoked previously can be applied using a generally more convenient transmission geometry. On the other hand, if the data collection has to be performed non-invasively on a bulk object, two main options are possible:

- either limiting the study to the object "surface" (as defined by the absorption of the incident beam and the reabsorption of the emitted beam) in a reflection geometry; or
- analysing the whole object using hard X-rays in a transmission geometry (*e.g.* tomography imaging).

The diversity of heritage materials that can be studied using synchrotron techniques is very high: metals, ceramics, glass, bones, teeth, textiles, polymers, etc., and their composites. The penetration depth of medium-energy X-rays is typically in the order of magnitude of 5–500 μm for a wide range of materials (see Table 2). The field of view is generally limited by the available translations of the sample stage, the overall congestion around the beamline nose and the time required for the data acquisition. For full-field techniques, the dimensions of the incident beam (usually a few square centimetres) are key parameters.

Table 2. Typical attenuation length (depth into the material where the intensity falls to $1/e$, ~37%, of its initial value) of X-rays at various energies in a few materials and percentage of the intensity of a 30-keV photon beam transmitted through a 1-cm block of matter. The ceramic composition is based on Pantos *et al.* (2002) and that of human hair on Bertrand (2002)

	Attenuation length			Transmitted intensity
Material	3 keV	10 keV	30 keV	after 1 cm at 30 keV
Human hair	48 μm	1.7 mm	27 mm	69%
Pure silica	7.9 μm	0.25 mm	5.9 mm	18%
Ceramic	7.8 μm	99 μm	2.2 mm	1.1%
Pure copper	1.5 μm	0.11 mm	0.11 mm	<1%

Using imaging techniques, the quantity of data generated can be exceedingly large, and its processing time is a factor that should not be underestimated.

3.1. Invasive experiments

The majority of heritage applications of synchrotron techniques rely on invasive experiments. Samples taken from heritage objects can then be analysed using a variety of techniques and, contrary to the common opinion, sampling is often not an issue, especially for the study of archaeological artefacts. The non-destructivity of synchrotron characterisation means that the sample can subsequently be reanalysed using complementary analysis techniques.

Optimum sample thickness has to be determined, such as (1) enough material is interacting along the beam path, (2) absorption of the output signal remains reasonable, and (3) the heterogeneity scale (that depends on the property studied) is compatible with the beam dimensions at the sample. The relative positioning of the sample and the detector can lead to parallax errors in reflection geometry, widening of the illuminated area, and degradation of the signal-to-noise ratio due to increased diffusion of X-rays.

For samples to be characterised in the infrared or X-ray range, no extra preparation than that involved in cutting it is usually required. Due to the penetration of X-rays, the surface state of the sample is far less crucial than for electron microscopy techniques, and the preparation protocols are far simpler than for most analytical methods. Conventional options are:

- directly cutting a compact sample using a scalpel blade, a saw, a micro-drill, etc., or using cryofracture protocols;
- embedding the sample in a resin medium, cutting, and thinning it with a polisher, if necessary; and
- using microtomy or cryomicrotomy techniques.

For experiments in the X-ray range, if the cross section has to be supported, ultra-thin polymer films are usually employed, such as low-absorbance amorphous Kapton© and high-purity Mylar© or Ultralene©. Mica windows are often used for SAXS experiments.

For infrared experiments, samples are usually deposited on a ZnS window that is transparent in the infrared region. Both films and embedding media have to be chosen according to the specific requirements of the method, and non-standard materials should ideally be tested before the actual experiment. Crystalline domains in polymer films can diffract and complicate the analysis of collected XRD diagrams. Films and embedding media sometimes contain significant contents of trace elements or can give rise to specific absorption features in the infrared domain. Up to now, no satisfactory preparation protocol has been reported in the literature enabling FTIR, XRF, and XRD experiments to be performed all-together on the same sample using a single preparation protocol.

Sophisticated sample environments such as furnaces, high-pressure diamond cells, chemical reactors, and electrochemical cells can be used at the beamline for *in situ* monitoring of firing processes, chemical reactions, corrosion mechanisms, etc., primarily on mock samples.

3.2. Non-invasive experiments

For museum objects, the analysis often needs not only to be non-destructive but also non-invasive. Additionally, samples previously prepared for other observation techniques (such as metallographic or painting cross sections for optical microscopy) may not possibly be reconditioned and have to be put in this category. The sample environment available at synchrotrons can vary a lot from one beamline to the other, and some setups may not allow the analysis of large-sized objects. Moreover, low-energy X-ray characterisations (typically for energies below 4 keV, *i.e.* calcium K absorption edge) require the object to be put under vacuum to avoid X-ray absorption by air. An alternative option around 4 keV is to blow a helium flow.

Due to the beam-characteristic attenuation length in the material, for most experiments, the information will come from a maximum depth of 5–500 µm into the object (see Table 2). This allows for a wide range of surface analyses on ceramics, glass, metal, paper, etc. A major limitation, working on bulk samples in reflection geometry, is that the signal resulting from the interaction between the beam and a heterogeneous material can be very hard to interpret. Even the layering of flat (quasi 2D) samples such as paintings and ceramic sherds is usually difficult to understand. For such samples, successful attempts show that insight into the layering can be derived from varying the incident beam angle or energy. An example using X-ray diffraction on ceramic samples can be found in Gliozzo *et al.* (2004).

An alternative, using hard X-ray beams available at synchrotron facilities (typically exceeding 30 keV), is to reconstruct 3D images using full-field tomography to collect density information. Using confocal configurations, methods such as XRF and energy-dispersive-XRD (Hall *et al.*, 1998) can also be employed to obtain 3D element and mineral-phase imaging inside a bulk object.

4. EXAMPLES OF IMAGING EXPERIMENTS

A very contrasted landscape prevails regarding the use of synchrotron techniques among the various heritage fields. Art and technology history is by far the one that has the most strongly used synchrotron imaging capabilities. On the contrary, the archaeometry community's need for elemental quantification essentially relies on methods such as ICP-MS, SEM-EDX, or IBA, mostly due to stronger pre-existing links between scientific communities. Recent trends tend to show an increase of the synchrotron imaging usage for archaeometry research. Compared to the number of works in those two fields, publications have scarcely dealt with conservation studies so far. Finally, paleoanthropological studies have developed specific synchrotron expertise for methods such as μCT. For the sake of clarity, experiments in this section are divided according to the scientific fields, however most of the approaches developed for a field could be transposed to any other.

4.1. Archaeometry

Spatial mapping of elemental content can be performed using μXRF on a variety of materials. Major, minor, and trace elements can all be observed and quantified, using appropriate processing and modelling of the data. The limit of detection is typically of the order of magnitude of 10^{-1}–1 part per million (ppm) in a typical glass matrix and 10^{-2}–10^{-1} in an organic matrix for transition metals (counting time: 1000 s, Somogyi *et al.*, 2001). The capability, specific to the synchrotron beam, to set accurate photon energy (monochromatisation) tremendously contributes to the study of samples for which quantification is difficult using laboratory XRF, PIXE, etc. One can excite the fluorescence to quantify minor or trace elements just below the absorption edge of major matrix constituents such as Pt in gold artefacts (Guerra *et al.*, 2005). Arsenic can be excited without stimulating Pb fluorescence, and some authors imaged As and Ni distributions in a late-Iron-Age sword that appears as a composite probably made by piling layers of distinct alloy compositions (Grolimund *et al.*, 2004).

Sciau *et al.* (2006) used a conjunction of μXRF for elemental composition determination and μXRD (both monochromatic and Laue) to identify the crystalline phases in the slip of Terra Sigillata ceramic pieces from La Graufesenque (Aveyron, France). Mineralogical maps were calculated through integration over given diffraction rings and contributed to the understanding of the peculiar hardness and visual appearance of this production.

Several authors have used the high monochromaticity and tunability of synchrotron radiation to raster scan samples at energies corresponding to pre-determined significant features of the absorption (XANES) spectrum. This enables the direct mapping of specific chemical (usually inorganic) compounds. Such an approach was used to map the oxidation states of iron in Roman smithing waste products. Interestingly mixed oxidation states, as in magnetite $Fe^{II}Fe^{III}_2O_4$ were mainly observed at phase boundaries (Grolimund *et al.*, 2004).

4.2. Art and technology history of museum objects

As indicated earlier, this area is the one that has most benefited from the development of synchrotron imaging techniques. A major reason is probably that some museum teams have developed a strong habit of collaborating with university laboratories or acquired in-house spectroscopy and diffraction equipment. In particular, the use of X-ray radiography had a strong impact in the field for the analysis of easel paintings and sculptures.

μXRF has been used for a variety of applications to obtain elemental maps on museum objects. An illustrative example is its use to contribute to the reading of the trace left after erosion of carved stone inscriptions through trace elemental 2D imaging (Bilderback *et al.*, 2005).

One of the most straightforward applications in the field comes from the identification of pigments in paint layers using μXRD (Salvadó *et al.*, 2002; Hochleitner *et al.*, 2003). Identification of pigments in painting cross sections is usually performed using a combination of optical microscopy and SEM-EDX. Such methods are nevertheless of limited utility to discriminate between pigments having similar colours and chemical compositions. It is as well difficult to correlate the information coming from the two microscopy methods. Salvadó *et al.* (2002) studied a painting by the Gothic Catalan painter Jaume Huguet (*ca.* 1415–1492) using synchrotron X-ray diffraction. The various mineral phases could be identified unequivocally and mapped across the painting stratigraphy. Results show that the pigment material comprises a large variety of crystalline compounds. Up to six different copper-containing phases were identified. Results were used by the authors to identify alteration compounds and to suggest putative pigment preparation protocols.

Non-invasive pigment identification can also be performed through 3D elemental analysis using confocal X-ray fluorescence imaging (Smit *et al.*, 2004) as was done by Kanngiesser *et al.* (2003) on an Indian Mughal miniature from the seventeenth to eighteenth century AD with a depth profiling resolution of *ca.* 10 μm. Areas of works several centimetres large can be entirely scanned using synchrotron techniques, as was carried out on a fragment of Roman mural painting using μXRD. The data obtained were processed using Rietveld refinement analysis to obtain a quantitative mineral-phase determination (Dooryhée *et al.*, 2005). However, one has to keep in mind that ill-crystallised or amorphous compounds may not be identified or even observed using XRD techniques.

Other examples include the μXRF mapping of the tin and mercury contents of an ancient mirror fragment. μXANES study at the Sn and Hg L_{III} edges confirmed that no oxidation of Hg had occurred in the ageing process (Bartoll *et al.*, 2004). The μXANES strategy described in the *Archaeometry* section was used to map Mn oxidation states in Roman glass fragments (Simionovici *et al.*, 2000).

4.3. Conservation and long-term corrosion

Stabilisation, cleaning, and consolidation treatments usually involve very complex chemistry. The study of the underlying mechanisms and the optimisation of their efficiency could probably benefit greatly from synchrotron radiation by better taking into account the local heterogeneity and texture complexity.

The study performed on the seventeenth-century warship *Vasa* (Sandström *et al.*, 2002) led to new results on the role played by iron species as catalysts to the degradation of waterlogged wood from marine archaeological environment through reaction with sulphur chemical groups. In a study on oak timber samples taken from the warship

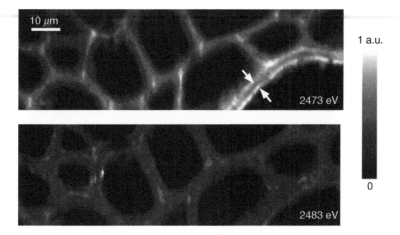

Fig. 5. μXRF map of an oak timber sample from the *Mary Rose* at 2473 and 2483 eV to image the distribution of reduced and oxidised sulphur species, respectively. Note the heterogeneous distribution of the reduced compounds in the wood-cell microstructure (*arrows*). Image courtesy of M. Sandström.

Mary Rose, Sandström *et al.* (2005) separately mapped reduced and oxidised sulphur species with ~0.5 μm resolution and correlated it with the iron distribution and wood micrometric structure (see Fig. 5, and Fors and Sandström, 2006).

Using μXRD, cross sections of parchment could be analysed in depth. The ordering of collagen molecules and lipids leads to specific features in the small angle diagrams collected. Results led to new information about parchment supramolecular structure and showed no effect of normal laser cleaning procedures on the parchment nanostructure, whereas partial gelatinisation was clearly observed for laser-burnt material (Kennedy *et al.*, 2004). The conservation of paper has also attracted attention recently, in particular regarding the degradation of iron gall inks (Janssens *et al.*, 2003; Kanngiesser *et al.*, 2004; Proost *et al.*, 2004).

Réguer *et al.* (2005, 2006) studied the corrosion processes of iron in samples from four twelfth to sixteenth century AD French archaeological sites. The distribution of chloride-containing phases plays a major role in the post-excavational degradation of iron artefacts. μXRF imaging (Cl, Fe) was used to visualise the elemental distribution. Subsequent local μXANES spectra in the regions of interest showed the correlation between Fe(III) akaganeite and low-Cl content, while confirming the presence of a newly observed β-Fe$_2^{II}$(OH)$_3$Cl phase in high-Cl regions (see Fig. 6). An early work showed maps of various iron species defined by their contributions to the Fe K-edge XANES spectrum (Nakai and Iida, 1992). The same strategy was used to study the corrosion mechanisms in the cross section of ancient bronze artefacts through the mapping of various oxidation states of copper (0, I, and II), confirming the presence of the Cu(II) brochantite (De Ryck *et al.*, 2003).

Fig. 6. (Top) Microscopy images showing the analysed area in a cross section of a corroded iron artefact from the fifteenth-century Avrilly archaeological site. (Bottom) µXRF maps at 2780 (just below the Cl K-edge) and 2830 eV (just above). The map at 2830 eV shows the heterogeneous chloride distribution among the various corrosion products. Images courtesy of S. Réguer.

4.4. Palaeontology and taphonomy

The interest of the palaeontology community for synchrotron µCT grew up very recently as a follow up to the development of laboratory µCT equipment. In the paleoanthropological field, 3D visualisation of individual teeth can be carried out non-destructively – thus avoiding the need to section invaluable samples to study their internal organisation (Chaimanee *et al.*, 2003; Tafforeau, 2004). The high contrast and signal-to-noise ratio, the parallel geometry, and the absence of *cupping*, due to the beam hardening effect using conventional X-ray sources, lead to an unprecedented quality of the reconstructions (see Section 3). Dental key characters (such as enamel thickness or roots morphology) were studied non-destructively with an accuracy never obtained before. These studies contribute to clarify phylogenetic issues about primates (see Fig. 7 and Marivaux *et al.*, 2006) and early hominids, including Tournai (*Sahelanthropus tchadensis*) (Chaimanee *et al.*, 2003; Brunet *et al.*, 2005; Tafforeau, 2004; Tafforeau *et al.*, 2006). 2D XRM techniques were also used to search centimetre-sized rock fragments looking for entrapped fossils (Ando *et al.*, 2000), to image insects in opaque amber (Tafforeau *et al.*, 2006) or to probe regions showing local alteration in archaeological bone samples (Wess *et al.*, 2002). The latter method could be developed as a wider screening strategy.

Keratinised tissues are usually degraded in burial environments except for very specific conditions of dryness (deserts), coldness (glaciers), and pH/humidity (marshes, with the so-called *bog bodies*). Studying such remnants is a unique chance to observe the long-term alteration of protein materials. Synchrotron µFTIR has been used in conjunction with laboratory FTIR to map the distribution of integral components of ancient hair and skin samples (lipids, proteins), to assess their preservation state, and to understand diagenesis mechanisms.

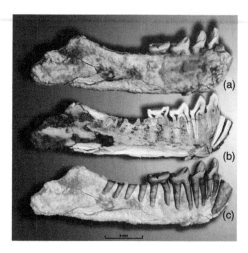

Fig. 7. Microtomographic investigation of the mandible of a newly identified Strepsirrhine primate from the Late Eocene of Thailand, *Muangthanhinius siami*. This experiment was performed on the beamline ID19 of the ESRF with a voxel size of 10.13 μm. (a) 3D rendering of the external morphology, (b) virtual sagittal section showing the internal anatomy, and (c) virtual pulling out of the teeth by 3D segmentation. Image courtesy of P. Tafforeau, L. Marivaux, Y. Chaimanee, and J.-J. Jaeger.

Such mechanisms most probably extend to the other keratin teguments such as wool. We used synchrotron FTIR on two Egyptian mummy hair fibres to map the oxidation process through the unravelling of keratin a-helices, using the secondary structure sensitivity of μFTIR for protein samples (Bertrand *et al.*, 2003). With the same technique, chemical group distributions were imaged in skin fragments from an Egyptian mummy to map the corresponding compounds (fatty acids, palmitic acid, calcium oxalate, etc.) and give clue to occurring diagenesis processes (Cotte *et al.*, 2005).

Using approaches similar to that developed by forensic sciences (for a review, see Kempson *et al.*, 2005), one can use synchrotron μXRF to look at the distribution of minor and trace elements in hair, teeth, and skin, and attempt to correlate their spatial distributions with the various putative sources. The trace elements in a Late Period mummy hair from Egypt could be attributed to surface contamination from dust, cosmetic treatments (Pb, Mn), and mummification protocols (see Fig. 8 and Bertrand *et al.*, 2003).

5. CONCLUSIONS: SOME WAY FORWARD

Comparing previous works to synchrotron analytical capabilities, one can try to identify some existing gaps and to derive possible developments that could be exploited in the near future.

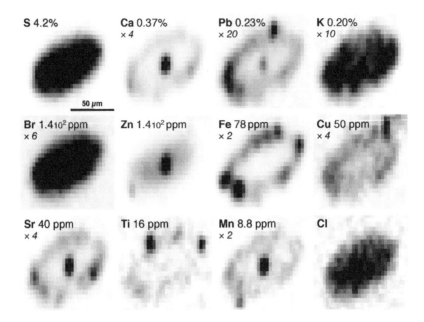

Fig. 8. Quantitative determination of major to trace elemental content in a late-period Egyptian mummy hair using μXRF. The italicised figures indicate the proportion relative to a standard human hair when significant differences are observed.

5.1. Complementary μXRF/μXRD/μXANES/μRaman

A current trend in Earth and environmental sciences is to develop complementary imaging techniques to map simultaneously or successively elemental, chemical, and structural information on the same sample. In particular, it could be very interesting to be able to determine the trace element content in specific mineral phases for provenance and technology studies. Trace element analysis could then be more precise than using the average content of a bulk sample without taking into account the various binding sites of the elements in the material structure.

Examples from other scientific fields such as environmental studies show that a fully quantitative approach could be adopted for μXANES, if necessary, leading to the quantitative mapping of individual chemical compounds on square-centimetre-wide areas (Pickering *et al.*, 2000). This could also be a step towards quantifying only trace elements in a given chemical environment for archaeometry studies.

5.2. Infrared Fourier-transform spectromicroscopy

Quite surprisingly, although FTIR spectroscopy is widely used in heritage laboratories, synchrotron infrared techniques do not seem to have attracted all the attention deserved from the community. The study of organic remnants, binding medium in paint layers, textiles, etc., should strongly benefit from the improved resolution and the mapping capability at synchrotron infrared beamlines.

5.3. Conservation

Many developments are expected in the conservation area to study the corrosion processes of a variety of materials, the binding of surface protective compounds, to follow the diffusion of consolidation products, PEGs, etc., in porous materials.

5.4. Organisation

For such developments to take place, a very strong collaboration needs to be established between the heritage community and synchrotron physicists. This also implies organisational consequences. For instance, in January 2004, the SOLEIL synchrotron (Gif-sur-Yvette, France) announced the setting up of a liaison office dedicated to cultural heritage research in response to the growing demand of the user community. SOLEIL later set up a specific review committee to select the proposals. This initiative is focussed on five major actions: (1) the building of a technical platform, (2) accessibility, (3) information and training activities, (4) valorisation, and (5) networking activities. Access to all beamlines of SOLEIL will be granted to the community, among which 6–8 are particularly expected to contribute to heritage research with an expected uppermost involvement of imaging beamlines (infrared, medium, and hard X-ray ranges) (Bertrand and Doucet, 2007; Bertrand *et al.*, 2006). The development of such interfaces at large-scale facilities is expected to contribute to the creation of new heritage-oriented technical clusters giving access to a variety of analysis methods at a supranational level.

ACKNOWLEDGEMENTS

The author wishes to thank particularly Drs. Jean Doucet (Laboratoire de physique des solides, Orsay, France), Emmanuel Pantos (CCLRC, Daresbury, UK), and Denis Raoux (Synchrotron SOLEIL) for their involvement in the development of the liaison office at SOLEIL. The author is grateful to Drs. Philippe Dillmann (Laboratoire Pierre Süe, Saclay France), Philippe Walter (Centre de recherche et de restauration des musées de France, Paris, France), Christian Degrigny (International Council of Museums, Committee for Conservation, Metal Working Group), Philippe Deblay (Synchrotron SOLEIL), and Prof. Annemie Adriaens (Ghent University, Belgium) for in-depth discussions about the project. The author would like to thank Drs. Magnus Sandström (Arrhenius Laboratory, Stockholm University, Sweden), Régis Bertholon (Université Paris I, France), Solenn Réguer (Laboratoire Pierre Süe, Saclay, France), Paul Tafforeau (European Synchrotron Radiation Facility, Grenoble, France) for authorising the publication of images from their work and for their helpful comments on the chapter. The author acknowledges the strong involvement of the SOLEIL synchrotron teams in the setting up of the liaison office. This work has been partly supported by a grant from the Région Île-de-France.

REFERENCES

Ando, M., Chen, J., Hyodo, K., Mori, K., Sugiyama, H., Xian, X., 2000. Nondestructive visual search for fossils in rock using X-ray interferometry imaging. *Jap. J. Appl. Phys.* **39**(2), L1009–L1011.

Bartoll, J., Röhrs, S., Erko, A., Firsov, A., Bjeoumikhov, A., Langhoff, N., 2004. Micro-X-ray absorption near edge structure spectroscopy investigations of baroque tin-amalgam mirrors at BESSY using a capillary focusing system. *Spectrochim. Acta B* **59**(10–11), 1587–1592.

Bertholon, R., 2001. Chapter Nettoyage et stabilisation de la corrosion par électrolyse. In *La conservation des métaux (Conservation du patrimoine Tome 5)*. Le cas des canons provenant de fouilles sous-marines, CNRS Éditions, Ministère de la Culture et de la Communication, Paris, France: pp. 83–101.

Bertrand, L., 2002. *Approche structurale et bioinorganique de la conservation de fibres kératinisées archéologiques.* PhD Thesis, Université Paris 6, Paris.

Bertrand, L., Doucet, J., 2007. Dedicated liaison office for cultural heritage at the SOLEIL synchrotron. *Nuovo Cimento* C. **30**(01), 35–40.

Bertrand, L., Doucet, J., Dumas, P., Simionovici, A., Tsoucaris, G., Walter, P., 2003. Microbeam synchrotron imaging of hairs from ancient Egyptian mummies. *J. Synchrotron Radiat.* **10**(5), 387–392.

Bertrand, L., Vantelon, D., Pantos, E., 2006. Novel interface for cultural heritage at SOLEIL. *Appl. Phys. A* **83**(2), 225–228.

Bilderback, D.H., Powers, J., Dimitrova, N., Huang, R., Smilgies, D.-M., Clinton, K., Thorne, R.E., 2005. X-ray fluorescence recovers writing from ancient inscriptions. *Z. Papyrol. Epigr.* **152**, 221–227.

Brunet, M., Guy, F., Pilbeam, D., Lieberman, D.E., Likius, A., Mackaye, H.T., Ponce de León, M.S., Zollikofer, C.P., Vignaud, P., 2005. New material of the earliest hominid from the Upper Miocene of Chad. *Nature* **434**(7034), 752–755.

Chaimanee, Y., Jolly, D., Benammi, M., Tafforeau, P., Duzer, D., Moussa, L., Jaeger. J.-J., 2003. A Middle Miocene hominoid from Thailand and orangutan origins. *Nature* **422**, 61–65.

Cotte, M., Walter, P., Tsoucaris, G., Dumas, P., 2005. Studying skin of an Egyptian mummy by infrared microscopy. *Vib. Spectrosc.* **38**(1–2), 159–167.

Davies, R.J., Burghammer, M., Riekel, C., 2005. Simultaneous microRaman and synchrotron radiation microdiffraction: Tools for materials characterization. *Appl. Phys. Lett.* **87**(26), 264105.

De Ryck, I., Adriaens, A., Pantos, E., Adams, F., 2003. A comparison of microbeam techniques for the analysis of corroded ancient bronze objects. *Analyst* **128**(8), 1104–1109.

Dooryhée, E., Anne, M., Bardiès, I., Hodeau, J.-L., Martinetto, P., Rondot, S., Salomon, J., Vaughan, G.B.M., Walter, P., 2005. Non-destructive synchrotron X-ray diffraction mapping of a Roman painting. *Appl. Phys. A* **81**(4), 663–667.

Fors, Y., Sandström, M., 2006. Sulfur and iron in shipwrecks cause conservation concerns. *Chem. Soc. Rev.* **35**, 399–415.

Gliozzo, E., Kirkman, I.W., Pantos, E. Memmi-Turbanti, I., 2004. Black gloss pottery: production sites and technology in Northern Etruria, part II: gloss technology. *Archaeometry* **46**(2), 227–246.

Grolimund, D., Senn, M., Trottmann, M., Janousch, M., Bonhoure, I., Scheidegger, A., Marcus, M., 2004. Shedding new light on historical metal samples using micro-focused synchrotron X-ray fluorescence and spectroscopy. *Spectrochim. Acta B* **59**, 1627–1635.

Guerra, M.F., Calligaro, T., Radtke, M., Reiche, I., Riesemeier, H., 2005. Fingerprinting ancient gold by measuring Pt with spatially resolved high energy Sy-XRF. *Nucl. Instrum. Methods B* **240**, 505–511.

Hall, C., Barnes, P., Cockcroft, J.K., Colston, S.L., Hausermann, D., Jacques, S.D.M., Jupe, A.C., Kunz, M., 1998. Synchrotron radiation energy-dispersive diffraction tomography. *Nucl. Instrum. Methods B* **140**, 253–257.

Hochleitner, B., Schreiner, M., Drakopoulos, M., Snigireva, I., Snigirev, A., 2003. Analysis of paint layers by light microscopy, scanning electron microscopy and synchrotron induced X-ray micro-diffraction. In *Proc. Conf. Art 2002*, Antwerp, Belgium, June 2003.

Isaure, M.P., Laboudigue, A., Manceau, A., Sarret, G., Tiffreau, C., Trocellier, P., Lamble, G., Hazemann, J.L., Chateigner, D., 2002. Quantitative Zn speciation in a contaminated dredged sediment by μ-PIXE, μ-SXRF, EXAFS spectroscopy and principal component analysis. *Geochim. Cosmochim. Acta* **66**(9), 1549–1567.

Janssens, K., Proost, K., Deraedt, I., Bulska, E., Wagner, B., Schreiner, M., 2003. Chapter, The use of focussed X-ray beams for non-destructive characterisation of historical materials. In *Molecular and Structural Archaeology: Cosmetic and Therapeutic Chemicals*, Volume 117. Kluwer Academic Publishers, Dordrecht, The Netherlands: pp. 193–200.

Kanngiesser, B., Malzer, W., Reiche, I., 2003. A new 3D micro X-ray fluorescence analysis set-up – first archaeometric applications. *Nucl. Instrum. Methods B* **211**, 259–264.

Kanngiesser, B., Hahn, O., Wilke, M., Nekat, B., Malzer, W., Erko, A., 2004. Investigation of oxidation and migration processes of inorganic compounds in ink-corroded manuscripts. *Spectrochim. Acta B* **59**, 1511–1516.

Kempson, I.M., Paul Kirkbride, K., Skinner, W.M., Coumbaros, J., 2005. Applications of synchrotron radiation in forensic trace evidence analysis. *Talanta* **67**, 286–303.

Kennedy, C.J., Hiller, J.C., Lammie, D., Drakopoulos, M., Vest, M., Cooper, M., Adderley W.P., Wess, T.J., 2004. Microfocus X-ray diffraction of historical parchment reveals variations in structural features through parchment cross sections. *Nano Lett.* **4**(8), 1373–1380.

LabsTech European Coordination Network (1999–2002). http://www.chm.unipg.it/chimgen/LabS-TECH.html

Manceau, A., Tamura, N., Marcus, M.A., MacDowell, A.A., Celestre, R.S., Sublett, R.E., Sposito, G., Padmore, H.A., 2002. Deciphering Ni sequestration in soil ferromanganese nodules by combining X-ray fluorescence, absorption, and diffraction at micrometer scales of resolution. *Am. Miner.* **87**(10), 1494–1499.

Marivaux, L., Chaimanee, Y., Tafforeau, P., Jaeger, J.-J., 2006. New Strepsirrhine primate from the late Eocene of peninsular Thailand (Krabi Basin). *Am. J. Phys. Anthrop.* **130**(4), 425–434.

Nakai, I., Iida, A., 1992. Applications of SR-XRF imaging and micro-XANES to meteorites, archaeological objects and animal tissues. In *Advances in X-ray Analysis*, Volume 35. Eds. Barrett, C.S., Gilfrich, J.V., Huang, T.C., Jenkins, R., McCarthy, G.J., Predecki, P.K., Ryon, R., Smith, D. Plenum Press: New York, USA, pp. 1307–1315.

Neff, D., Réguer, S., Bellot-Gurlet, L., Dillmann, P., Bertholon, R., 2004. Structural characterization of corrosion products on archaeological iron: an integrated analytical approach to establish corrosion forms. *J. Raman Spectrom.* **35**(8–9), 739–745.

Pantos, E., Tang, C.C., MacLean, E.J., Cheung, K.C., Strange, R.W., Rizkallah, P.J., Papiz, M.Z., Colston, S.L., Roberts, M.A., Murphy, B.M., Collins, S.P., Clark, D.T., Tobin, M.J., Zhilin, M., Prag, K., Prag, A.J.N.W., 2002. Applications of synchrotron radiation to archaeological ceramics. In *Modern Trends in Scientific Studies on Ancient Ceramics*, Volume 1011. Eds. Kilikoglou, V., Hein, A., Maniatis, Y. BAR International Series, Oxford, United-Kingdom: pp. 377–384.

Pickering, I.J., Prince, R.C., Salt, D.E., George, G.N., 2000. Quantitative, chemically specific imaging of selenium transformation in plants. *PNAS* **97**(20), 10717–10722.

Proost, K., Janssens, K., Wagner, B., Bulska, E., Schreiner, M., 2004. Determination of localized Fe^{2+}/Fe^{3+} ratios in inks of historic documents by means of μ-XANES. *Nucl. Instrum. Methods B* **213**, 723–728.

Réguer, S., Dillmann, P., Mirambet, F., Bellot-Gurlet, L., 2005. Local and structural characterisation of chlorinated phases formed on ferrous archaeological artefacts by μXRD and μXANES. *Nucl. Instrum. Methods B* **240**(1–2), 500–504.

Réguer, S., Dillmann, P., Mirambet, F., Susini, J., Lagarde, P., 2006. Investigation of Cl corrosion products of iron archaeological artefacts using micro-focused synchrotron X-ray absorption spectroscopy. *Appl. Phys. A* **83**(2), 189–193.

Salvadó, N., Pradell, T., Pantos, E., Papiz, M.Z., Molera, J., Seco, M., Vendrell-Saz, M., 2002. Identification of copper-based green pigments in Jaume Huguet's Gothic altarpieces by Fourier transform infrared microspectroscopy and synchrotron radiation X-ray diffraction. *J. Synchrotron Radiat.* **9**(4), 215–222.

Salvo, L., Cloetens, P., Maire, E., Zabler, S., Blandin, J.J., Buffière, J.Y., Ludwig, W., Boller, E., Bellet, D., Josserond, C., 2003. X-ray micro-tomography an attractive characterisation technique in materials science. *Nucl. Instrum. Methods B* **200**, 273–286.

Sandström, M., Jalilehvand, F., Persson, I., Gelius, U., Frank, P., Hall-Roth, I., 2002. Deterioration of the seventeenth-century warship *Vasa* by internal formation of sulphuric acid. *Nature* **415**(6874), 893–897.

Sandström, M., Jalilehvand, F., Damian, E., Fors, Y., Gelius, U., Jones, M., Salomé, M., 2005. Sulfur accumulation in the timbers of King Henry VIII's warship *Mary Rose*: a pathway in the sulfur cycle of conservation concern. *PNAS* **102**(40), 14165–14170.

Sciau, P., Goudeau, P., Tamura, N., Dooryhée, E., 2006. Micro scanning X-ray diffraction study of Gallo-Roman Terra Sigillata ceramics. *Appl. Phys. A* **83**(2), 219–224.

Simionovici, A., Janssens, K., Rindby, A., Snigireva, I., Snigirev, A., 2000. Precision micro-XANES of Mn in corroded Roman glasses. In *X-ray Microscopy: Proceedings of the VI International Conference*, Berkeley, California, USA, 2–6 Aug 1999. Volume 507 of *AIP Conference*, Eds. Meyer-Ilse, W., Warwick, T. and Atwood, D. American Institute of Physics: Melville, New York, USA, pp. 279–283.

Šmit, Ž., Janssens, K., Proost, K., Langus, I., 2004. Confocal μ-XRF depth analysis of paint layers. *Nucl. Instrum. Methods B* **219–220**, 35–40.

Somogyi, A., Drakopoulos, M., Vincze, L., Vekemans, B., Camerani, C., Janssens, K., Snigirev, A., Adams, F., 2001. ID18F: a new micro-x-ray fluorescence end-station at the European Synchrotron Radiation Facility (ESRF): preliminary results. *X-Ray Spectrom.* **30**(4), 242–252.

Tafforeau, P., 2004. *Aspects phylogénétiques et fonctionnels de la microstructure de l'émail dentaire et de la structure tridimensionnelle des molaires chez les primates fossiles et actuels: apports de la microtomographie à rayonnement X synchrotron.* PhD Thesis, Université de Montpellier II, France.

Tafforeau, P., Boistel, R., Boller, E., Bravin, A., Brunet, M., Chaimanee, Y., Cloetens, P., Feist, M., Hoszowska, J., Jaeger, J.J., Kay, R.F., Lazzari, V., Marivaux, L., Nel, A., Nemoz, C., Thibault, X., Vignaud, P., Zabler, S., 2006. Applications of X-ray synchrotron microtomography for non-destructive 3D studies of paleontological specimens. *Appl. Phys. A* **83**(2), 195–202.

Wess, T.J., Alberts, I., Hiller, J., Drakopoulos, M., Chamberlain, A.T., Collins, M., 2002. Microfocus small angle X-ray scattering reveals structural features in archaeological bone samples: detection of changes in bone mineral habit and size. *Calcif. Tissue Int.* **70**(2), 103–110.

Chapter 3

Holistic Modeling of Gas and Aerosol Deposition and the Degradation of Cultural Objects

I.S. Cole, D.A. Paterson and D. Lau

CSIRO Manufacturing & Infrastructure Technology, PO Box 56, Highett, Victoria 3190, Australia
Email: Ivan.Cole@csiro.au

ho·lis·tic (hō-lĭs'tĭk) adj. Emphasizing the importance of the whole and the interdependence of its parts.
Concerned with wholes rather than analysis or separation into parts.

Abstract

This chapter addresses the deposition of gases and aerosols both inside and outside museums and the possible effects that such deposition may have on cultural objects. This issue is addressed through the concept of holistic modeling, where all critical factors controlling the deposition and degradation process are defined and linked together. The types and sizes of particulates both within and exterior to a museum are outlined. The types of gases found within a dwelling and their relations to exterior pollutants are described. The aerosol and gas deposition mechanisms and the equations for each mechanism are outlined. In order to define conditions for gas deposition, the factors controlling condensation and formation of moisture layers are also presented. These principles and equations are then illustrated by analysis of the generation, transport and deposition of aerosols on cultural objects in the external environment, followed by a similar analysis for inside buildings. In the case of deposition inside buildings, the literature is first reviewed, and then three case studies are analyzed that represent significant cases or highlight unresolved issues in the literature. The case studies clarify the relative importance of each deposition mechanism. It is evident that the major mechanisms within a building are gravity, vortex shedding and, in case of significant air flows, momentum-dominated impact. Factors controlling the attachment and detachment of pollutants both within and outside dwellings are then outlined, as are the common damage forms that result for some pollutants. Throughout the chapter and especially towards the end, the implications of the findings to design and maintenance strategies are discussed.

Keywords: Particulates, gases, deposition, condensation, cultural objects, interior spaces, pollutant transport, degradation, maintenance.

Contents

1. INTRODUCTION

This chapter addresses the deposition of liquid and solid particulates and gases on cultural objects that may be open to the external environment or protected inside buildings, and the effects of such pollutants and condensates.

In an external environment, the major pollutants are marine aerosols and acidic particulates, which may promote corrosion (oxidation), lead to damage from crystallization and promote condensation, providing a culture medium for biodegradation. Within buildings (museums or historic structures), there may be a wide range of particulates and pollutants (see Section 2). These may be damaging to collections in a number of ways. They may:

- act as a carrier for more degradation species;
- lead to visual but not chemical degradation;
- chemically degrade artwork; and
- act as a catalyst for chemical degradation.

Where benign particulates deposit and do not chemically interact, damage may be induced secondarily when treatments for removal lead to the ingress of deposited material into the surface structure of an object. This may occur with dry-surface-cleaning treatments involving brushing or rubbing, or wet surface swabbing or washing.

A further consideration is the susceptibility of the object to the deposited particulate or gas. Inorganic materials may be relatively unaffected by microbial deposits, and acidic deposits are generally considered aggressive for most materials, while co-deposited gases or particulates may act to neutralize an acidic or alkaline pollutant.

Baer and Banks (1985) indicate that soot and organic compounds may absorb damaging gases such as SO_2. Soot may also lead to visual degradation of paintings. $(NH_4)_2SO_4$ can induce bloom on varnish, while other S-rich materials can be responsible for discoloring of the pigments by oxidation to H_2SO_4 (De Santis et al., 1992). This process can be catalyzed by Fe-rich particles.

There have been a range of studies looking at pollutant sources, transport and deposition, both in the exterior and interior environments, and these are reviewed in Sections 4.1, 5 and 6.1. One limitation of many of these studies is that they tend to look at a particular component of the problem of degradation, rather than study the series of processes that lead to it. Degradation is an interconnected process involving pollutant generation, transport and deposition, followed by attack on the surface and physico-chemical modification of the surface. The current authors have developed a holistic model of degradation that links and models the processes controlling atmospheric corrosion on a range of scales, from macro through meso to local, micro and lastly micron (Cole et al., 2003). A schematic presentation of the holistic model as defined by the event sequence and scale diagram is given in Fig. 1.

These scales are defined in line with EOTA (1997), so that "macro" refers to gross meteorological conditions (polar, subtropical, etc.), "meso" refers to regions with dimensions up to 100 km, "local" is in the immediate vicinity of a building, while "micro" refers to the absolute proximity of a material surface. As indicated above, the holistic model links

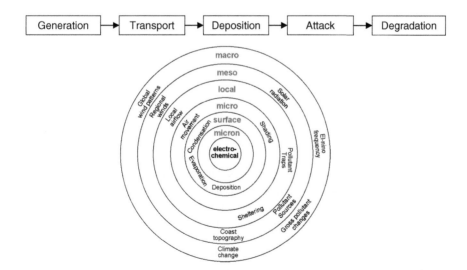

Fig. 1. Framework for a holistic model. A holistic view of the time sequence, of the environment and of degradation mechanisms.

processes controlling degradation across a range of scales, so that the formation of marine aerosols is modeled on the meso-scale from wave activity in the oceans, while oxide growth is modeled on the micron scale. The emphasis of the model shifts when modeling exterior or interior conditions.

In the case of the exterior of a building, the emphasis of the research is on determining the aerosol concentration adjacent to a dwelling (which involves significant source and transport modeling). Deposition onto a dwelling is presented, but can be modeled relatively simply as it is primarily controlled by turbulent diffusion (Cole and Paterson, 2004). In contrast, deposition within a building can occur by a variety of mechanisms (Camuffo, 1998), of which turbulent diffusion is one of the least important. The pollutant concentration within a building is controlled by exterior pollutant levels and by *ad hoc* factors within a building. Thus, in this chapter, the discussion of internal deposition will focus on the mechanism of pollutant deposition.

This chapter will address the types of particulates and gases, deposition mechanisms, deposition equations, generic case studies for cultural objects outside and inside buildings, attachment and detachment issues, surface forms of corrosion and degradation, and implications for design and maintenance.

2. TYPES AND SIZES OF PARTICLES AND TYPES OF GASES

2.1. Particulates

In this chapter, the word "particle" is used for both solid and liquid aerosols, but not for gaseous pollutants. Common dangerous classes of particulate deposits are (Hill and Bouwmeester, 1994):

- acidic substances,
- oxidizing substances,
- soot and tarry particles from burning fuel,
- large abrasive particles,
- hygroscopic materials,
- particles containing traces of metals (that act as catalysts),
- wet and oily particles from food preparation,
- alkaline particles from new concrete,
- salt crystals and dissolved salts (corrosion and microorganisms),
- textile fibers and skin fragments as food for insects.

Particulates may arise from exterior sources, either natural (sea salt aerosol, bushfires, windblown dust, pollen and other plant products and insect, arachnid and bird products), agricultural (agricultural sprays), transport (oil and soot) or industrial (organics and sulfuric acid).

Particles may also arise from within, either human-related (lint, dirt, hair, skin flakes, droplets and food), microorganism-related (mites, mold and other microorganisms), combustion-related (cigarette smoke, soot and ash from candles, incense, smoke from the kitchen range and frying oil), renovation-related (brick, plaster and asbestos dust and cleaning products) or resuspended (from vacuuming, walking, dusting and sitting on furniture).

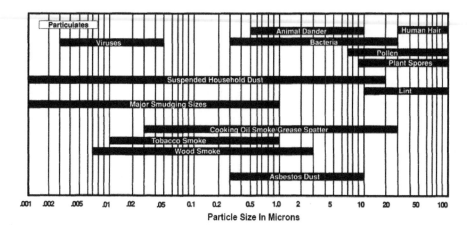

Fig. 2. Size ranges of indoor pollution particles.

In a study of particulate soiling in museums and historic houses (Yoon and Brimblecombe, 2001), sticky samplers were used to collect particulates. The main components were soil dust, soot and fibers with lesser amounts of human hair and skin, plant and paint fragments and insect parts. Some typical particle size ranges are given in Fig. 2 (Annis, 1991).

2.2. Gases

In the external environment of urban regions, the major acidic and oxidizing gaseous pollutants occur as a result of industrial activity. Sulfur dioxide and nitrogen dioxide are produced from the combustion of fossil fuels (coal, gas and oil) used in energy production and transport. While natural sources may be the primary contributors for these pollutants in remote areas, human activity is by far the major source in the modern built environment. Although naturally occurring in the stratosphere, ozone, an oxidizing pollutant, is found at ground levels and is most often a result of photochemical smog. Levels of ozone can be directly related to those of SO_x and NO_x.

The consequences of outdoor materials are (1) acidic and (2) oxidizing. Specific implications of acidic attack exist for calcareous materials, cellulosic materials and particularly ferrous but most metals. Buildings and monuments made of limestone, marble and other alkaline stones are literally dissolved through the neutralizing reaction with acids.

Oxidizing attack occurs through destructive oxidation of chemical bonds within organic materials, *e.g.* synthetic polymers, natural polymers (cellulose and protein) and dyes. However, the short half-life of ozone somewhat modifies its powerful oxidizing action.

Acidic activity of any particular acid is indicated by its dissociation constant pKa (see Table 1). The overall acidic effect experienced by a surface is contributed to by the sum of all the atmospheric acids present, including carbonic (from CO_2) and organic acids, but is influenced by pKa. Therefore, considering that nitric and sulfuric acids have a pKa much

Table 1. pKa values for acids at 25°C in water (CRC Handbook of Chemistry and Physics, Edition 76, 1995–1996)

Acid	Formula	pKa
Perchloric	$HClO_4$	~−7
Hydrochloric	HCl	~−7
Chloric	$HClO_3$	~−3
Sulfuric (1)	H_2SO_4	~−2
Nitric	HNO_3	~−1.3
Oxalic (1)	$H_2C_2O_4$	1.23
Sulfuric (2)	HSO_4^-	1.92
Chlorous	$HClO_2$	1.96
Phosphoric (1)	H_3PO_4	2.12
Nitrous	HNO_2	3.34
Formic	$HCOOH$	3.75
Oxalic (2)	$HC_2O_4^-$	4.19
Acetic	CH_3COOH	4.75
Carbonic (1)	H_2CO_3	6.37
Hydrosulfuric	H_2S	7.04
Ammonium ion	NH_4^+	9.25
Hydrogen peroxide	H_2O_2	11.62

(1) and (2) refer to the first and second deprotonation stages of the acid.

lower than carbonic acid, in a solution where the concentration of nitric and sulfuric acids may be much lower, these highly acidic species will contribute much more to the total acidity.

While particulates are generally considered to have adverse effects for both humans and objects, the definition of pollutant gases for collections remains more open to interpretation and reliant on the particular situation. Gases that have measurable health risks for humans are not always dangerous for collections, and the reverse also applies. Those most often associated with harm to collections are SO_x, NO_x, carboxylic acids (acetic, formic and fatty acids), ozone and some amines. Volatile organic compounds (VOCs) are included as they are often present in association with the short-chain carboxylic acids and their precursors, although the specific effects of most VOCs are still unknown. Larger molecules may photodissociate to produce active smaller molecules or be involved in inter- or intramolecular chemical reactions. For example, VOCs with primary alcohol groups may react with oxygen to become aldehydes, and then carboxylic acids. Gases can also interact and influence the particulate composition of an environment. Ozone and terpenes have been shown to increase the number and mass concentrations of sub-micron particles, resulting in an undocumented source of indoor particulates (Weschler and Shields, 1999).

It is also worth noting that in studies of concentrations of indoor air VOCs, total VOC (TVOC) concentrations are considerably higher than individually measured VOCs (Brown *et al.*, 1994). This suggests that there are a large number of chemical compounds present, but only a select few are evaluated using current monitoring methodologies.

In the inside museum environment, gaseous pollutants arise from both internal and external sources. Internally generated VOCs are generally recognized to occur in far greater concentration indoors compared with the external environment, in the ratio 8:1 for established buildings and greater than 200:1 for new buildings (Brown, 2003). Conversely, pollutants generated outdoors are found to be in much lower concentration indoors, as exemplified by the measurements of SO_2, O_3, NO_x and NO given in Table 2.

Gaseous emissions may occur from the artworks themselves, or from the materials and coatings used in storage, transport and display. Processed wood products are the most often cited sources of VOCs such as acetate and formate. Newer building materials are expected to reduce emissions to an acceptable level with time; however, the persistence of emissions is demonstrated in older wood in a study by Rhyl-Svendsen and Glastrup (2002). It was found that a 15-year-old oak plank had an acetic acid emission of 55.7 (±5.6) µg/m²h and a formic acid emission of 11.1 (±2.5) µg/m²h; and a 12-year-old coin collection drawer with a Masonite board base and maple wood edges had an acetic acid emission of 172.5 (±6.9) µg/m²h and a formic acid emission of 94.1 (±7.2) µg/m²h.

3. DEPOSITION MECHANISMS

3.1. Aerosol deposition mechanisms

There are a number of mechanisms that can lead to the deposition of particulates (Camuffo, 1998; Van Greiken et al., 1998). In this chapter, the most important of these are restated, some with new or improved equations, and a few more possible deposition mechanisms are introduced.

The main deposition mechanisms for aerosol particles are:

- gravitational settling,
- turbulent diffusion,
- laminar diffusion (also called Brownian deposition and diffusiophoresis),
- thermophoresis (migration from high temperatures to low),
- electrostatic attraction,
- momentum-dominated impact,
- vortex shedding (transport by transient laminar flows),
- filtering (flow past or through raised fabrics) and
- photophoresis (motion generated by an intense beam of light).

These are illustrated in Fig. 3.

Different particle sizes are affected by different mechanisms. Gravity and momentum-dominated impact are most significant for the largest particles. Turbulent diffusion and vortex shedding are independent of particle size. Electrostatic attraction and laminar diffusion are strongest for the smallest particles. Thermophoresis and filtering display a weak dependence on particle size, with smaller particles being favored in thermophoresis.

Because the effect of vortex shedding on deposition is not widely known, it is described here in a bit more detail than the others. It is similar to turbulent diffusion, but occurs at much

Table 2. Ratio of indoor–outdoor gaseous pollutants

Ref.	Location	SO_2	O_3	NO_x	NO
Brimblecombe (1990)	National Gallery, London		0.003–0.002		
Brimblecombe (1990)	Sainsbury Centre, Norwich		0.7		
Brimblecombe (1990)	Baxter Gallery, Pasadena		0.6		
Brimblecombe (1990)	Scott Gallery, California		0.45	0.92	1.02
Brimblecombe (1990)	Huntington Library			0.94	1
Brimblecombe (1990)	Rijksmuseum, Amsterdam	0.1–0.2			
Brimblecombe (1990)	Summer				
Brimblecombe (1990)	Rijksarchief, Den Haag	<0.25	<0.03	0.35	
Brimblecombe (1990)	Rijksarchief, Arnhem		0.6	0.22	0.6
Brimblecombe (1990)	Rijksarchief, Leewwarden			<0.15	2
Brimblecombe (1990)	Winter				
Brimblecombe (1990)	Rijksarchief, Den Haag	<0.1		0.1	1.1
Brimblecombe (1990)	Rijksarchief, Arnhem	0.65		0.32	1.2
Brimblecombe (1990)	Rijksarchief, Leewwarden	0.04		<0.1	1.4
Drakou (1998)	Laboratory of Atmos Physics, Thessaloniki, Greece		0.61	0.8	0.61
Drakou (1998)	Laboratory of Atmos Physics, Athens		0.38	0.63	0.63
Salmon et al. (2000)	Wawel Castle, Poland		0.19		
Salmon et al. (2000)	Wawel Castle, Poland		0.17		
Salmon et al. (2000)	Matejko Museum, Poland		0.43		
Salmon et al. (2000)	National Museum, Poland		0.23		
Salmon et al. (2000)	National Museum, Poland		0.19		
Salmon et al. (2000)	Collegium Maius, Poland		0.24		
Salmon et al. (2000)	Cloth Hall Museum, Poland		0.44		

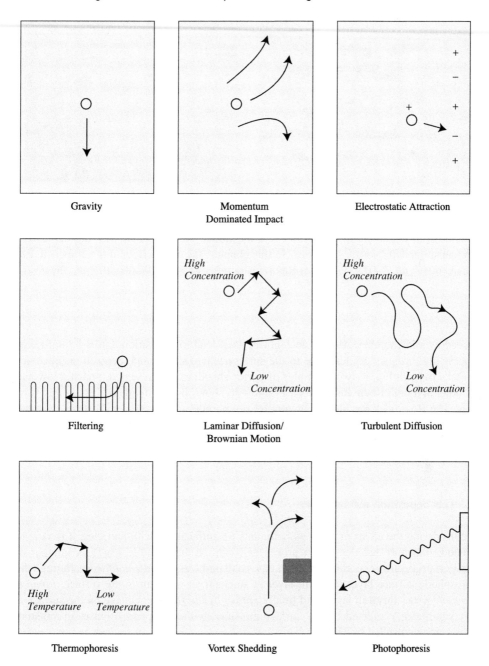

Fig. 3. Schematic diagram of deposition mechanisms.

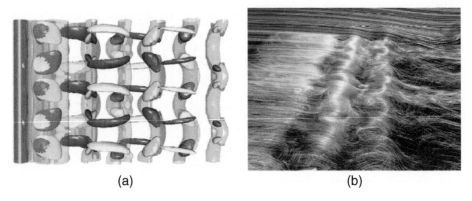

(a)   (b)

Fig. 4. Examples of the chaotic transient flow associated with airflow around obstacles. Flow is from left to right. It is flows like these that are associated with the deposition mechanism called "vortex shedding" in this chapter: (a) with a Reynolds number of 190 (Blackburn *et al.*, 2005); (b) with a Reynolds number of 3000 (Weinkauf *et al.*, 2003).

lower Reynolds number. Also, it does not in itself lead to deposition, but is a long-range process that brings particles close to the surface where short-range deposition processes can take over. In Fig. 4, two examples of vortex shedding are presented, one at a low and the other at a high Reynolds number. Figure 4(a) shows both individual vortices and flow lines. The flow lines are blue or green and run parallel to the surface. Vortices (red and yellow) exist both within and between the flow lines. Figure 4(b) (Weinkuaf *et al.*, 2003) shows the streamlines that result when transitional flow occurs around a backward-facing step. Two shear layers roll up into vortices.

3.2. Gas deposition mechanisms

In considering the effect of gaseous pollutants on surfaces, as with particles, it is vital to consider transport and deposition processes (Seinfeld and Pandis, 1997). A number of transport processes are common to both gases and particles: *turbulent diffusion* through the atmospheric surface layer to a thin layer of stagnant air adjacent to the surface; *laminar diffusion* across this thin layer; and finally *uptake by the surface*. Uptake by the surface is very significantly affected by any surface moisture, with absorption of gases in moisture films being controlled by Henry's law

$$A(g) \Leftrightarrow A(aq) \tag{1}$$

where

$$[A(aq)] = H_A p_A \tag{2}$$

Table 3. Relevant properties of gases for dry deposition calculations (extracted from Seinfeld and Pandis (1997))

Species	H_A^* (M atm^{-1}) at 298 K	Normalized reactivity
Ozone	1×10^{-2}	1
Nitrogen dioxide	1×10^{-2}	0.1
Hydrogen sulfide	0.12	–
Ammonia	2×10^4	0
Sulfur dioxide	1×10^5	0
Hydrogen peroxide	1×10^5	1
Nitrate radical	2.1×10^5	–
Hydrochloric acid	2.05×10^6	0

*Effective H_A assuming a pH of 6.5.

where p_A is the partial pressure of A in the gas phase (atm), $[A(aq)]$ is the aqueous phase concentration of A (mol l^{-1}) and H_A is Henry's law coefficient.

The surface absorption of gaseous species is controlled by the normalized reactivity. Table 3 presents the effective Henry's law coefficient H_A and normalized reactivity values for various gas species. The effective H_A in Table 3 takes into account reactions of the aqueous species.

When H_A is high and the normalized reactivity is low (*e.g.* sulfur dioxide), deposition will be primarily through absorption into a moisture droplet (be it a wet aerosol or a surface moisture film). When H_A is low and the normalized reactivity is high (ozone), direct gaseous absorption will dominate.

4. DEPOSITION EQUATIONS

4.1. Aerosol deposition equations

The overall deposition velocity v can be calculated from the deposition velocities for the individual mechanism outlined above, depending on whether the mechanisms act in parallel or in series. When the mechanisms occur together or in parallel (at the same distance from the surface), the deposition velocities can simply be added together:

$$v = \sum_i v_i \tag{3}$$

If the deposition mechanisms occur in different layers and the depositing aerosol must pass through all layers (act in series), deposition velocities must match between layers:

$$v_i = v_j \tag{4}$$

A typical example (using vortex shedding with laminar diffusion) of the way in which deposition velocities are matched between layers is by automatic adjustment of $(\nabla\phi)_2$ in the near-surface layer to force equality, as in:

$$v_{ld} = v_{vs} = \frac{\Gamma_{vs}(\nabla\phi)_1}{\phi} = \frac{D(\nabla\phi)_2}{\phi} \tag{5}$$

The next sections define the equations for deposition velocity for each mechanism.

In deriving deposition velocities, it is first necessary to define some factors that quantify "paths" through air. Air can be treated as a continuous medium on the large scale and as a mixture of moving molecules on the sub-micron scale. This is handled by introducing the Knudsen number and the Cunningham slip factor. The Knudsen number is:

$$Kn = \frac{2\lambda}{D_p} \tag{6}$$

where $\lambda = 6.51 \times 10^{-8}$m is the molecular mean free path in air, and D_p is the particle diameter. The Cunningham slip factor C_c is a slip correction factor for particles that are small relative to the molecular mean free path in air. Many formulae have been derived for this. One of the best is (Seinfeld and Pandis, 1997):

$$C_c = 1 + Kn\left[1.257 + 0.4\exp\left(-\frac{1.1}{Kn}\right)\right] \tag{7}$$

4.1.1. Gravitational settling

Gravity is the main mechanism for the removal of particles from air. The settling rate is governed by a balance between the downward gravitational force and the aerodynamic drag. When the air velocity is small, as inside a building, the settling velocity relative to the air velocity can be written approximately as:

$$v_t = \frac{D_p^2 \rho_p g C_c}{18\mu(1 + Re_t / 128)} \tag{8}$$

This formula is derived in-house by combining formulae from Seinfeld and Pandis (1997) and Clift *et al.* (1978).

The terminal velocity due to gravity v_t depends on the particle diameter D_p, density of a single particle ρ_p, acceleration due to gravity = 9.81 m/s^2, slip correction factor C_c and air viscosity $\mu = 1.789 \times 10^{-5}$ Pa s.

The Reynolds number is:

$$Re_t = \frac{v_t D_p \rho_a}{\mu} \tag{9}$$

where $\rho_a = 1.225$ kg/m^3 is the density of air. This becomes important when the particle diameter exceeds 100 μm.

The largest airborne particles are seldom spherical (because spherical particles settle too fast to become airborne). This is taken into account by numerically reducing the particle density to get the equivalent terminal velocity (Clift *et al.*, 1978; Seinfeld and Pandis, 1997). Hence, for sand and silt, $\rho_p = 2100$ kg/m³; for water-based particles (including bacteria and cooking oil fumes), $\rho_p = 1200$ kg/m³; for soot and clay flakes, $\rho_p = 500$ kg/m³; and for lint, $\rho_p = 100$ kg/m³.

When the air velocity is not small, such as outside a building, the air drag vector becomes more complicated and the simple formula of equation (3) breaks down completely. Then, the only way to handle gravitational settling is by the use of computational fluid dynamics (CFD). This has been done.

4.1.2. Momentum-dominated impact

This is the dominant deposition mechanism for large particles on vertical surfaces outside buildings.

The best way to calculate momentum-dominated impact is to use CFD, and this has been done for external surfaces.

For surfaces inside buildings, the lower airspeeds mean that momentum-dominated impact is of less importance. It can occur indoors where there are air jets caused by sneezing, blowing of noses, opening and closing of doors, rapid hand movements near walls and thermal plumes above heaters. There are many equations in the literature (Laitone, 1979; Seinfeld and Pandis, 1997), but they all seem to require a knowledge of upstream conditions, which is not available in practice. For that reason, a new equation was developed in-house:

$$v_i = -\frac{x_0}{\tau} + \sqrt{\left(\frac{x_0}{\tau}\right)^2 + U_0^2} \tag{10}$$

where U_0 is the approach velocity at distance x_0 from the wall and the relaxation time τ is given by (Seinfeld and Pandis, 1997).

$$\tau = \frac{D_p^2 \rho_p C_c}{18\mu} \tag{11}$$

It is to be stressed that a simple formula like equation (10) is not as good as the use of CFD.

4.1.3. Laminar diffusion/Brownian deposition

The movement of air molecules can provide enough force to move sub-micron particles around. This diffuses clouds of particles and leads to a net diffusion from regions of high concentration to those of low concentration. If particles are depositing on a surface, then this deposition locally lowers the particle concentration near the wall, and other particles diffuse in to take their place. This deposition mechanism has been described in detail by Seinfeld and Pandis (1997) and Camuffo (1998).

The deposition velocity for laminar diffusion is (Camuffo, 1998):

$$v_{ld} = \frac{D\nabla\phi}{\phi} \tag{12}$$

Equation (12) is calculated from diffusivity, particle concentration ϕ and particle concentration gradient $\nabla\phi$. This deposition velocity is sufficiently slow that no correction for Reynolds number is needed.

From Camuffo (1998) and Seinfeld and Pandis (1997), the Brownian diffusivity D is given by:

$$D = \frac{kTC_c}{3\pi\mu D_p} \tag{13}$$

Laminar diffusion is only significant inside buildings. Outside buildings, the deposition is greatest when the wind is strongest and, at those times, the wind is invariably turbulent, and turbulent diffusion completely dominates over laminar diffusion.

4.1.4. Turbulent diffusion

Turbulent diffusion moves particles from regions of high concentration to those of low concentration. Outside buildings, it is the dominant mechanism that lifts particles against the force of gravity, because gravity creates a concentration gradient. Inside buildings, it is negligible.

Turbulent diffusion differs from laminar diffusion in several important respects. First, it is almost independent of particle size. Second, it moves particles to about 10 µm from surfaces where some other deposition mechanism (usually momentum-dominated impact) takes over.

Where the flow is parallel to the surface and is fast enough for turbulence, the deposition velocity is given by:

$$v_{td} = \frac{\Gamma_t \nabla\phi}{\phi} \tag{14}$$

Turbulent diffusion outside buildings is best calculated using CFD. The diffusion coefficient is calculated for every point in space from parameters of the turbulence model from an equation like:

$$\Gamma_t = \frac{0.09k^2}{\sigma_t \varepsilon} \tag{15}$$

where k and ε are the turbulent kinetic energy and turbulent energy dissipation, respectively, and $\sigma_t = 0.9$ is the turbulent Prandtl number.

4.1.5. Vortex shedding

Vortex shedding is a time-dependent periodic forced convection that occurs when an airflow encounters an obstacle. Particle transport is enhanced as the Reynolds number is increased, and the flow becomes more chaotic. Deposition due to vortex shedding is more significant inside buildings than outside, but is included in both analyses.

The deposition velocity is given by (derived in-house):

$$v_{vs} = \frac{\Gamma_{vs} \nabla \phi}{\phi} \tag{16}$$

The following *ad hoc* equation for the diffusion coefficient was developed in-house for internal deposition from data by Sohankar *et al.* (1995), Sheard *et al.* (2005) and Blackburn *et al.* (2005):

$$\Gamma_{vs} \approx 0.05 V d\, f(Re) \tag{17}$$

where $f(Re)$ is a function of the Reynolds number that allows for incomplete action between Reynolds number of 53 and 260, *e.g.*:

$$f(Re) = \begin{cases} 0.25 & \text{for} & 53 \le Re < 190 \\ 0.5 & \text{for} & 190 \le Re < 260 \\ 1 & \text{for} & 260 \le Re \end{cases} \tag{18}$$

4.1.6. Filtering

When air flows through or over soft furnishings, the fibers of the soft furnishing can filter large particles from the air. The mechanisms associated with filtering are described in detail by Annis (1991), but no equations are given in that publication.

It is the airflow parallel to the surface of the fabric that is analyzed in this chapter. The flow through fabric from one side to the other is sufficiently specialized to ignore here, and would need to be treated separately on a case-by-case basis.

Because of the general lack of soft furnishings outside buildings, filtering is only important inside buildings.

The deposition velocity is given by (derived in-house):

$$v_f = \frac{\Gamma_f \nabla \phi}{\phi} \tag{19}$$

The effect of flow past the fibers of a fabric can be modeled in several ways. If turbulent diffusion is significant, then the formula for turbulent diffusion is adjusted to:

$$\Gamma_t = \frac{u\kappa(y + y_0)}{\sigma_t} \tag{20}$$

where y_0 is the mean distance between fiber bundles.

If the local flow is laminar, then diffusion due to filtering is:

$$\Gamma_f \approx \frac{0.25 y_0^2 V}{y} \tag{21}$$

4.1.7. Electrostatic attraction

This can be important for small particles inside buildings. It is important near surfaces, rather than far a way. Many aerosols are nearly electrically neutral – exceptions are soot and ash produced by combustion.

The amount of electrostatic attraction of surfaces relates to their electrical conductivity and water-holding ability. For example, materials like nylon, polythene, polystyrene, varnish and epoxy are said to hold electrostatic charges well. The deposition velocity of charged particles is roughly given by (derived in-house):

$$v_e = \frac{qC_c E}{3\pi\mu D_p (1 + Re_e / 128)} \tag{22}$$

where q is the charge on the particle and E is the electric field strength generated by the surface. The Reynolds number is $Re_e = v_e D_p \rho_a / \mu$.

The electric field near a uniformly charged insulating surface is:

$$E = \frac{V}{d} \tag{23}$$

where V is the voltage and d is the distance from the surface at which the voltage is measured (Tuma, 1976).

If the wall has regions of positive and negative charges separated by a mean distance r_σ, then the electric field is well approximated by:

$$\frac{2\varepsilon_0 E}{\sigma_w} = \begin{cases} 1 & \text{for} \quad r/r_\sigma \leq 0.15 \\ (2.8 - r/r_\sigma)/(2.8 - 0.15) & \text{for} \quad 0.15 < r/r_\sigma \leq 2.5 \\ 10^{1.5(1.9 - r/r_\sigma)} & \text{for} \quad 2.5 < r/r_\sigma \end{cases} \tag{24}$$

It is essentially zero for $r \geq 4\ r_\sigma$.

The electrostatic charge may be generated by chemical reactions (*e.g.* via pH), by thermionic emission or by friction.

Thermionic emission affects soot and ash, which are produced at high temperatures, and they lose electrons in this process. The steady state charge at any given temperature is given by the Saha equation (Prado and Howard, 1978) together with an equation for the diffusion of electrons in air.

Friction is the main mechanism whereby carpets, curtains and soft furnishings become electrostatically charged. The voltage is produced by friction with clothing, boots and other soft furnishings. There are no reliable equations, so it is essential to use experimental data. Woolen carpets charged by rubber-soled shoes will pick up a temporary charge of more than +20 000 V. Silk rugs can pick up about +10 000 V. Polyester picks up a small charge on the order of –700 V. Rubber (shoes and carpet backing) can reach a negative temporary charge of perhaps –50 000 to –100 000 V.

4.1.8. Thermophoresis

Thermophoresis is the migration of particles from regions of high temperature to those of low temperature. The deposition rate due to thermophoresis is almost independent of particle size for aerosols that are thermal insulators, but it decreases with increasing particle size for particles that are good thermal conductors. Thermosphoresis is only significant inside buildings.

There are many equations in the literature, but one of the more accurate ones is that of Seinfeld and Pandis (1997), *i.e.*:

$$v_{Th} = -\frac{3C_c\mu(k_a + c_t k_p Kn)\nabla T}{2\rho_a T(1 + 3c_m Kn)(k_p + 2k_a + 2c_t k_p Kn)} \tag{25}$$

where T is temperature (K), ∇T is the temperature gradient, k_a and k_p are air and particle thermal conductivities, respectively, and $c_t(=2.2)$ and $c_m(=1.0)$ are constants.

4.1.9. Photophoresis

Photophoresis has not been modeled in the present study. It could be modeled using the equations of Camuffo (1998) and Cadle (1965) or by the equations of Goldman (1978).

4.2. Pollutant deposition equations

The state of the surface (presence of moisture layers) is critical in modeling the deposition of gaseous pollutants, and in this section, therefore, equations controlling the formation of condensation layers are presented.

4.2.1. Condensation

The presence of a moisture layer is a prime requirement for the corrosion of metals, the deposition of gaseous pollutants and the growth of microbiological agents in almost all circumstances. As a first approximation, a surface will become wet when the surface relative humidity (RH_s) is greater than the deliquescence RH of any contaminating salts. RH_s can be calculated from the ambient air RH and the surface temperature of the metal. Work by Cole *et al.* (2004d) has validated this approach. In exterior environments, the surface temperature of an exposed metal can be derived by considering both undercooling to the night sky and daytime solar heating. For undercooling, the rate of heat transfer between the surface of any object and the sky through radiation is given by:

$$\frac{Q}{A} = \varepsilon\sigma(T_{sky}^4 - T^4) = h_r(T_a - T) \tag{26}$$

where Q is the heat flow rate (in watts), A is the surface area facing the sky, ε is emissivity, σ is the Stefan–Boltzmann constant ($=5.6697 \times 10^{-8}$ W m^{-2} K^{-4}), T_{sky} is the mean temperature of the sky, T is the temperature of the surface of the object, h_r is radiation heat transfer coefficient and T_a is the local air temperature.

However, estimation of surface temperatures also needs to take into account the effect of convection (air above object) and conduction (within the object), and thus wind speed becomes an important limiting factor (Cole *et al.*, 2005a). Clouds also tend to reduce undercooling (due to reflection of radiation from the earth). Change in the surface temperature of an object in the morning depends on the change in local air temperature, direct and diffuse solar radiation on the object and heat loss from evaporation. Further, if the surface is wet (*e.g.* from condensation), a number of other factors apply. Firstly, the effect of evaporation and condensation on heat transfer to the surface must be considered. Secondly, emissivity of water ($\varepsilon = 0.96$) rather than of galvanized metal ($\varepsilon = 0.12$) is used in calculating the radiative heat losses. Thus, undercooling will be significantly reduced by the presence of a condensation layer. After sunrise, the rise in the plate temperature will also be significantly reduced by the cooling effects of evaporation from the surface. Taking all these factors into account, an accurate estimation of surface temperature can be derived (Cole and Paterson, 2006).

4.2.2. Gas deposition

The uptake of gaseous species into moisture layers is governed by Henry's law (see Section 3.2), but may also be significantly affected by aqueous reactions, as once the gases are dissolved, the hydrolyzed products may dissociate (Seinfeld and Pandis, 1997):

$$SO_2 \cdot H_2O \leftrightarrow H^+ + HSO_3^- \tag{27}$$

$$\text{while } HSO_3^- \leftrightarrow H^+ + SO_3^{2-} \tag{28}$$

The balance of HSO_3^- versus SO_3^{2-} depends on pH, with the former dominating at pH values of 2.5–7, and the latter above this. A major aqueous reaction of importance is the oxidation of SO_3^{2-} or other forms of S(IV) to SO_4^{2-} or other forms of S(VI). This may occur via a variety of mechanisms, including reactions with O_3, H_2O_2 and O_2 (catalyzed by Mn(II), Fe(III) and NO_2):

$$S(IV) + 0.5\ O_2 \xrightarrow{Mn^{2+},\ Fe^{3+}} S(IV) \tag{29}$$

For example, where S(IV) may be HSO_3^- versus SO_3^{2-} and S(VI) is HSO_4^{2-} or SO_4^{2-}.

According to Hoffman and Calvert (1985), the reaction rate is given by:

$$1.2 \times 106\ [Fe(III)][S(IV)] \tag{30}$$

The sulfate that forms from the oxidation of S(IV) may exist as SO_4^{2-} or HSO_4^- (H_2SO_4 dissociates to HSO_4^- under pH ranges likely for droplets in the atmosphere or on surfaces) according to:

$$HSO_4^- \rightleftharpoons H^+ + SO_4^{2-} \tag{31}$$

However, the absorption of ammonia into moisture films will tend to reduce the acidity produced by SO_2 absorption through the following equations:

$$NH_3 + H_2O \rightleftharpoons NH_3 \cdot H_2O \tag{32}$$

$$NH_3 \cdot H_2O \rightleftharpoons NH_4^+ + OH^- \tag{33}$$

Similar to SO_2, the formaldehyde solubility in water is greater than that indicated by H_A alone, as it readily hydrolyzes to methylene glycol according to

$$HCHO_{(aq)} + H_2O \leftrightarrow H_2C(OH)_{2(aq)} \tag{34}$$

Since the total amount of dissolved formaldehyde T(HCHO) will be the sum of dissolved formaldehyde and methylene glycol, and the concentration of the two are related by the law of mass action,

$$K = \frac{\left[H_2C(OH)_{2(aq)} \right]}{\left[HCHO_{(aq)} \right]} \tag{35}$$

where K is the equilibrium constant (mol/l) for the hydrolysis described in equation (34), then T(HCHO) may be expressed as

$$T(HCHO) = H_A \, pHCHO(1+K) \tag{36}$$

5. GENERATION, TRANSPORT AND DEPOSITION ON CULTURAL OBJECTS EXPOSED TO THE EXTERNAL ENVIRONMENT

The particulates that most affect degradation of cultural objects exposed to the outside environment are plant seeds, fungal spores, particulates produced by industrial and transport sources, and marine aerosols.

Within the corrosion science literature, there are a number of studies that provide either empirical evidence or models of marine aerosol distribution over land. The emphasis in this chapter is on marine aerosols, as air pollutants are limited and decreasing in the areas under study. Ohba et al. (1990) model the salt concentration in the air over land and divide salt into two components – salt produced at the shoreline, which shows an exponential decrease with distance from the coast, and a second component that relates to salt produced at sea, which is relatively constant. They relate this difference to the coarser nature of surf-generated salt compared to ocean-generated salt. Gustafsson and Franzen (1996) carried out an extensive monitoring program on Sweden's west coast and found that dry deposition of sea salt depended on the wind velocity on the coast and the downward distance from the coast. In the field of corrosion science, several surveys (Johnson and Stanners, 1981; Strekalov and Panchenko, 1994; Corvo et al., 1995, 1997; Cole and Ganther, 1996) have been undertaken in different countries to study the variation of salinity on land with distance from the coast. In reviewing some of this literature, Morcillo et al. (1999) found that, in general, a double dependence was evident, with a rapid decrease within the first few hundred meters and then a slower decrease tending towards an symptotic value. However, significant variations occurred in the dependence between the regions analyzed. Earlier, Strekalov and Panchenko (1994), on the basis of measurements of salinity in Murmansk and Vladivostok (Russia), found that the deposition of chlorides depended on both the average velocity of total winds (marine and continental) and on the product of wind velocity and duration, which they referred to as wind power. Morcillo *et al.* (2000) used data derived from Spain's Mediterranean coast to significantly extend Strekalov and Panchenko's concept of wind power, and proposed that certain marine wind directions (which they referred to as saline winds) are critical to the deposition of marine salts across land.

Cole *et al.* (2003) have integrated marine aerosol generation and transport into the holistic model of corrosion. Marine aerosols are produced by breaking waves, both on the shoreline and in the open ocean. In the open ocean, aerosols are produced by whitecaps of ocean waves. Whitecap production varies systematically with longitude and season, being at a maximum in low latitudes in July and at a maximum in high latitudes in December, and low all year round in tropical seas. Thus, tropical seas produce a relatively low volume of marine aerosol, resulting in decreased marine corrosion in near-equatorial regions. Salt production is also controlled by ocean effects such as local wind speed, beach slope and fetch.

Factors controlling the transport of particulates of all types are outlined by Cole *et al.* (2003). Aerosol residence times are controlled by convection and gravity, and aerosol scavenging by cloud drops, raindrops and physical objects on the ground (trees, buildings, etc.). Thus, marine aerosol transport is likely to be higher in dry climates with low rainfall and low ground coverage, while it will be restricted in humid and high-rainfall climates with forest cover. Aerosols produced by surf tend to be coarse (5–20 μm), and those produced by whitecaps are generally smaller (0.5–3 μm).

Thus, surf-produced aerosol rapidly deposits (due to gravity), while ocean-produced aerosol may be transported over considerable distances.

An Australia-wide map of airborne salinity derived in this way (Cole *et al.*, 2004a) is presented in Fig. 5. Coastal salinity depends on latitude, and the salinity inland depends on the distance from the coast. The salinity map highlights the pronounced effect of both ocean state (as defined by whitecap activity) and climate factors in controlling airborne salinity in Australia. For instance, southern Australian coastal zones, where whitecap activity is high, have appreciably higher airborne salinity levels than Australia's northern coast, where whitecap activity is low.

Aerosol deposition onto exposed objects has been computed using CFD and the equations presented in Section 4. It is primarily controlled by gravity, momentum-dominated impact and turbulent diffusion (Cole and Paterson, 2004). Figure 6 shows the competing influences of momentum-dominated impact and turbulent diffusion on a cylinder of arbitrary diameter at arbitrary wind speed. From Fig. 6, it is evident that turbulence diffusion dominates deposition for small particles ($D_p < 0.5$ μm) and that the turbulence intensity has a dominating influence on the deposition efficiency. At larger diameters, particle diameter is dominant, and deposition efficiency rises strongly with particle diameter.

The size and shape of objects are important. For complex forms such as cultural objects, deposition efficiency will vary across a structure, with deposition being highest at the

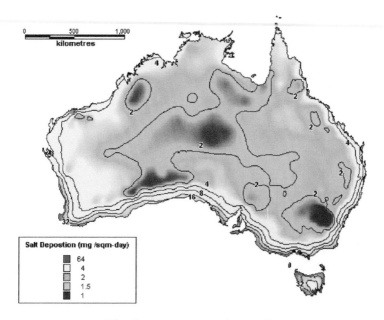

Fig. 5. Salinity map of Australia.

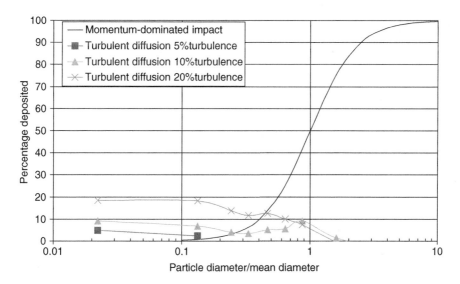

Fig. 6. Influences of momentum-dominated impact and turbulent diffusion on the deposition of aerosols on a cylinder of arbitrary diameter at arbitrary wind speed, as a percentage of the aerosol flux that would pass through in the absence of the cylinder. The percent turbulence is the turbulence intensity (rms velocity/mean velocity) upstream.

Fig. 7. Aerosol deposition on a building 10 m high and 20 × 20 m in plan. The flow is from right to left. Blue is low concentration, and red is high concentration.

edges of a structure where turbulence is highest. Figure 7 presents the results of a CFD simulation in which particles impact on a simplified building of dimensions 10 m height and 20 m by 20 m in plan. About 4 million particles were released in an area of 982 m^2 upstream. There are 10 colors ranging from blue to red. Red corresponds to 18% or more of the upstream mass flow density, and blue corresponds to 2% or less. The aerosol diameter is 10 μm.

5.1. Implication from source transport and deposition models to degradation of objects

A knowledge of airborne salinity and its deposition is, of course, relevant to the preservation of metal objects exposed in the open. The mapping work defined in Fig. 5 provides an indication of environmental severity in a given geographical location. For example, where salinity is greater than 8 mg/m^2·day, moderate atmospheric corrosion (of course, the extent of the corrosion rate depends on the material) is probable. However, preventative maintenance, primarily regular washing, would be sufficient to protect most metal work (with the exception of uncoated iron or steel). Where salinity is greater than 32 mg/m^2·day, severe corrosion is possible for some metals, and more active protective measures such as coatings may be required.

Local factors may, however, change these effects. The effects relate to roughness and turbulence. Surface features, trees, buildings, etc., scavenge salt from the atmosphere, depleting aerosol concentration to roughly the height of the object. Thus, a structure in a landscape with significant features of greater height than itself will have a reduced salt load compared to the same structure in an open area. This can of course be used to one's advantage in landscaping, as trees may be planted to reduce salt loads on structures that are upwind from marine breezes. Structures that rise above the surrounding landscape may show a maximum in salt deposition (and thus possibly corrosion) just above the height of surrounding objects. The second major effect is that of turbulence. This can, of course, lead to differential deposition on an object (as in Fig. 6), and thus additional attention during maintenance should be given to edges on objects. However, turbulence may also affect objects placed on structures. For example, "gargoyles" placed on medieval churches are at a position of high turbulence and will thus suffer from increased pollutant and aerosol deposition. In a more modern context, care should be taken when placing structures near or around sculptures or other openly exposed pieces to ensure that they do not produce heightened turbulence on the cultural object. This may be negated to some extent due to moisture effects, where increased turbulence will promote drying and lead to decreased corrosion rates, which is especially important when hygroscopic salts are present on a surface. However, increased dryness will also reduce degradation due to biological activity from molds and mildew.

6. GENERATION, TRANSPORT AND DEPOSITION ON CULTURAL OBJECTS INSIDE BUILDINGS

6.1. Literature review

Surveys have indicated particular patterns in particulate deposition within museums. Van Greiken et al. (2000) reported on studies in three museums: the Correr Museum and Kunsthistorisches Museum in Italy, and the Sainsbury Centre for Visual Arts in the United Kingdom.

In the Correr Museum, degradation of wall plaster constitutes an important source of indoor particulate matter rich in Ca. In the Kunsthistorisches Museum, particles originating from construction works can enter the museum through the air-conditioning system. In the Sainsbury Centre, practically all the indoor particles originated from outdoors. Particulates of sea salt are observed in each museum, but at lower concentrations than in the exterior, while the concentrations of aluminosilicates is comparable on the interior to the exterior.

Gysels et al. (2004) report on a study of the Koninklijk Museum voor Schone Kunsten (KMSK) in Antwerp, Belgium. The most common particulates were NaCl and soil-derived particulates (e.g. Al, Si and Ti). Soil-derived particulates were primarily in the coarse size range (>10 µm), while Na and S- and Fe-rich particulates were in the fine size range (<2.5 µm). Concentrations of NO_2 and SO_2 were evaluated at 12 and 5–6 ppb, respectively, independent of the season.

Camuffo et al. (2001) report on conditions in the KMSK, as well as the three museums mentioned above. They report that, in the Kunsthistorisches Museum, large particles (diameter >4 μm) are only observed near the floor level. Camuffo *et al.* (2001) noted an interesting diurnal cycle in which these coarse particles tended to be resuspended during the day and redeposited at night.

Injuk *et al.* (2002) studied the composition and size of particulates found in the Miyagi Museum of Art in Sendai, Japan. They measured the particulate matter in the museum restaurant and in two galleries. The most common particulates were AlSi and organic particulates, with the AlSi being dominant in the restaurant. The organic particulate size was generally >16 μm, while the most common dimension for AlSi, Si-rich and K-rich particulates was in the 4–16 μm range. Calcium sulfates, T-rich and Zn-rich particles were most common in the 2–4 μm range, while Fe-rich particulates were most common in the 0.5–1 μm range. S-rich particulates were most common in the 0.5–1 μm in the restaurant, but in the 4–16 μm range in the galleries.

The field observations confirm the list of particulates in the previous section and high-light the role of coarse aluminosilicates, finer marine and S-rich particulates and particu-lates with an indoor source (organics or Ca-rich particulates). The survey also indicates the role of air-conditioning/heating systems, humans and cooking in circulating (and promoting deposition of particulates) or as a source of particulates.

Studies in the Uffizi show higher levels of SO_2, NO_2 and O_3 outdoors compared with indoors, but higher levels of HONO were encountered indoors, suggesting a mechanism whereby indoor generation exists (Katsanos *et al.*, 1999), and a mechanism for produc-tion induced through heterogenous reactions relying on surface roughness is proposed.

The effectiveness of many deposition mechanisms is dependent on the wind speeds within museums. Albero *et al.* (2004) conducted a CFD study of airflow in the gilded vault hall in the Domus Aurea in Rome. They predicted airspeeds of up to 0.28 and 0.2 m/s asso-ciated with doors and roof vents, respectively. Papakonstantinou *et al.* (2000) modeled the airflow in the main hall of the National Archeological Museum of Athens. In general, airflows were estimated as quite moderate (0–0.3 m/s); however, higher speeds were estimated close to the door (up to 4 m/s) and at the ceiling level (up to 1.1 m/s). Papakonstantinou *et al.* (2000) associated the higher speeds at the ceiling level to local air heating due to lighting.

Some work in the general field of indoor air quality is relevant to the deposition of particulates in museums. Thatcher *et al.* (2002) experimentally determined the deposition of particles (0.5–10 μm) in an isolated room as a function of room furnishings and air speed. Increasing the airspeed from >0.05 to 0.19 m/s increased the deposition rate for all particle sizes, while increasing the amount of furnishing caused significant increases in deposition, particularly for fine particles.

6.2. Issue arising from the literature

The literature indicates that three types of particulates may be either prevalent of particu-larly damaging, *i.e.*:

- large aluminosilicate particulates, which are a major constituent of particulate matter and, although not chemically damaging, may lead to soiling;

- soot that may absorb damaging chemical species and may deposit due to electrostatic attraction;
- fine particulates such as S-rich particulates and sea salt that may be chemically damaging.

Further, the literature indicates that the following processes may promote deposition:

- ventilation-induced flow, which may particularly affect objects on walls,
- airflow from open doors or windows and
- airflow induced by heating or cooking.

6.3. Case studies

With the above literature in mind, the following three case studies have been selected in order to investigate the whole range of internal deposition phenomena. All the equations from Section 4 have been evaluated for each.

1. *Settling on upward-facing surfaces*: air speed 0.01 m/s at 0.15 m from the surface; no temperature difference; obstacles (furniture, furniture legs, etc.) near ground level have a typical width of 0.1 m.
2. *Flow past a vertical surface at 0.2 m/s at a distance of 0.05 m from surface*: temperature difference 1.5°C (cold surface); projections from the surface (picture frames, window frames, door frames, etc.) have a typical width of 0.03 m.
3. *Aerosol deposition on downward-facing surfaces*: Case 3(a) – kitchen, air speed 1 m/s at a distance of 0.1 m from the surface, temperature difference 10°C (cold surface); Case 3(b) – hall, air speed 0.2 m/s at a distance of 0.15 m from the surface, temperature difference 2.5°C (cold surface).

Some constants applicable to the case studies are given in Table 4.

Table 4. Parameters and values used in case studies

Parameter	Material	Symbol	Value
Air viscosity		μ	1.789×10^{-5} Pa s
Air density		ρ_a	1.225 kg/m^3
Mean free molecular path length	Air	Λ	6.51×10^{-8} m
Boltzman constant		k	1.38×10^{-23} J/K
Air temperature		T	288 K
Aerosol particle density	Sand/silt	ρ_p	2100 kg/m^3
	Water-based, including bacteria and cooking-oil smoke		1200 kg/m^3
	Soot, clay flakes		500 kg/m^3
	Lint		100 kg/m^3

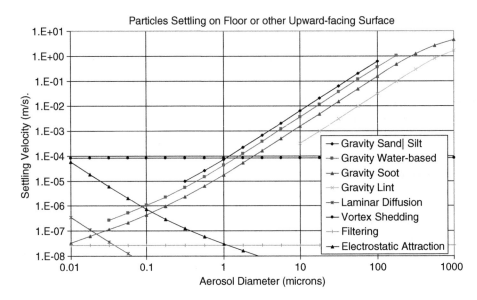

Fig. 8. Settling velocities on upward-facing surfaces as a function of aerosol particle diameter for a number of different deposition mechanisms.

6.3.1. Case study 1 – settling on upward-facing surfaces

The deposition or settling velocity is given in Fig. 8.

Gravity is analyzed for the four types of particulates in Table 3. However, the curve for sand/silt is taken to include asbestos fibers, and that for water-based is taken to include microorganisms and cooking-oil smoke.

From Fig. 8, it is evident that, for particles with the transition from vortex shedding to gravity as the dominant deposition mechanism, as the particle size increases, the settling velocity increases dramatically (by three orders of magnitude as the size increases from 1 to 100 μm).

Vortex shedding only brings particles down to about 1 mm from the surface. For that final 1 mm, the particle is brought down to the surface by laminar diffusion, impact, gravity, electrostatic effects or filtering. Filtering only brings it from 1 mm away down to about 10 μm from the surface. Below 1 mm, filtering and laminar diffusion are much more efficient than would be expected from Fig. 8 (up to 150 times stronger than shown).

The electrostatic charge on deposited particles is small and only affects the deposition of smaller particles.

6.3.2. Case study 2 – flow past a vertical surface

The deposition or settling velocity is given in Fig. 9. Vortex shedding from window, door and picture frames is by far the most dominant mechanism for deposition near these obstacles. Similar processes such as the moving of doors and drafts from open windows have similar deposition rates as this.

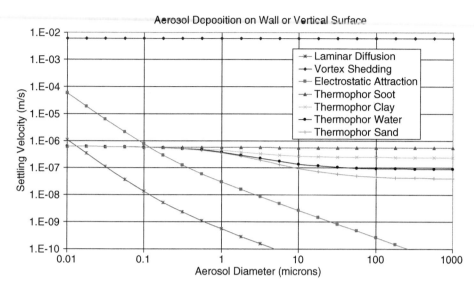

Fig. 9. Settling velocities on vertical surfaces as a function of aerosol particle diameter for a number of different deposition mechanisms.

Away from these objects, thermophoresis dominates. Electrostatic attraction may be significant for small particles, particularly for curtains.

6.3.3. Case study 3 – deposition on a downward-facing surface

Figure 10 presents the settling velocities for a rising air jet above a cooking appliance in a kitchen, and Fig. 11 presents the settling velocities for a rising jet of air above a radiator (or due to forced convection from air passing through an open window or door) in a hall.

In both cases, the competition between momentum-dominated impact and gravity results in a saw-tooth dependence of settling velocity on the particle size. As the particle size increases, the efficiency of momentum-dominated impact as a deposition mechanism increases. However, when a critical particle size is reached, gravity causes the particles to fall down and not deposit.

It is notable that, despite the large temperature differences that have been modeled for a ceiling, the effect of thermophoresis is only of marginal importance. Likewise, the effects of vortex shedding on ceilings is smaller than that for walls, because the fittings that induce periodic or chaotic flow are smaller in diameter on the ceiling (typically 1 cm across) and are oriented perpendicular to the ceiling, rather than parallel as on the walls.

6.4. Discussion of case studies

Following the review of the literature, a series of questions were posed (Section 6.2). These can be paraphrased as: How do large aluminosilicate, soot and finer (but chemically damaging)

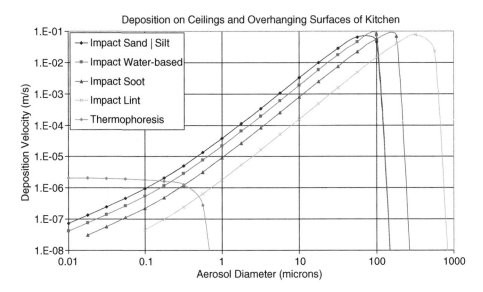

Fig. 10. Settling velocities on vertical surfaces in a kitchen as a function of aerosol particle diameter for a number of different deposition mechanisms. Both the impact and thermophoresis curves have had the effect of gravity subtracted from them.

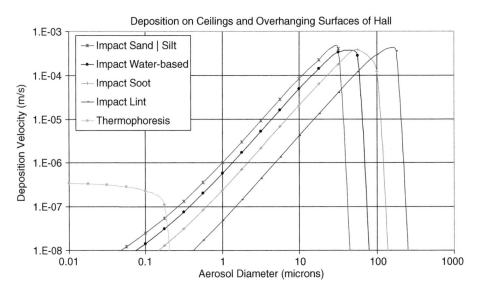

Fig. 11. Settling velocities on vertical surfaces in a hall as a function of aerosol particle diameter for a number of different deposition mechanisms. Both the impact and thermophoresis curves have had the effect of gravity subtracted from them.

S-rich and marine-derived particulates deposit? What is the role of ventilation-induced flow and airflow from open doors, windows, heating or cooking on deposition?

For all particles, the dominant mechanism is highly dependent on particle size. For large particles (>100 µm) such as aluminosilicates, gravity is dominant and rapid (close to 1 m/s), so that such particles will tend to deposit on upward-facing surfaces at entry points (floors near doorways, etc.).

Medium-size particles such as finer aluminosilicates and coarser S-rich and marine aerosols (1–100 µm) will deposit both by gravity and as a result of momentum-dominated impact in airstreams with moderate to high velocities. These airstreams may arise from thermal plumes (*e.g.* above a cooking surface or a radiator) or due to breezes from opening and closing doors, downwind of badly placed ventilators and due to human movement (hand movement, sneezes, etc.). Both mechanisms have settling velocities of the same order of magnitude, and thus in a "typical" museum, medium-size particles will deposit both on upward-facing surfaces and surrounding airstreams throughout the building. Deposition in the surrounding airstreams is highly dependent on the speed of the airstream – particles of 100 µm will deposit at a speed of 0.1 m/s in a strong thermal plume above a cooking surface, but only at about 4×10^{-4} m/s in a weaker plume above a radiator.

For fine particles, including the majority of S-rich and marine aerosols, the strongest deposition mechanism is due to forced convection periodic and chaotic laminar flows, lumped here under the title "vortex shedding." For thermal jets parallel to walls, interaction with fittings can lead to deposition velocities of up to 0.006 m/s across the whole range of particle diameters. In more sedately moving airflows, this drops to about 10^{-4} m/s near obstacles. Vortex shedding only brings particles down to about 1 mm from the surface and relies on other deposition mechanisms to complete the deposition process. Thus, fine and chemically damaging particulates may deposit on objects placed on walls if the air-conditioning or heating system generates air circulatory patterns with moderate air speeds (0.1 m/s). In the absence of such air circulation, deposition will be much slower with contribution from gravity, impact-dominated momentum and through electrostatic attraction of lightly charged particles, leading to a distribution of particles throughout the museum.

6.5. Attachment and detachment

6.5.1. Within buildings

Resuspension of aerosols from horizontal surfaces is a major issue. This is particularly the case for chemically inactive and uncharged particles, as particles that are attached by strong chemical bonds and by physical wrapping (*e.g.* evaporates) can be impossible to resuspend. Walking on carpets moves particles above the 1-mm layer where they can be convected upwards by forced and free convection processes. Walking produces local airspeeds higher than 1 m/s. The use of soft underlays can enhance the resuspension by walking. Sitting on soft furnishings ejects particles high into the convection zone, as does dusting, while sweeping moves particles above the 1-mm layer like walking, and also produces local airspeeds above 1 m/s. Vacuuming is the worst of all for resuspension.

Small particles (possibly those <5 μm in diameter) pass through the paper bag filters. Hypo-allergenice vacuum cleaners claim to stop most 0.3 μm particles. Camuffo *et al.* (2001) attributed the resuspension to air turbulence generated by visitors and the heating, ventilating and air conditioning system (HVAC) system. They also indicated that air movement promoted by HVAC systems may increase the deposition of airborne particles. In fact, museums with a high number of visitors and carpets showed a ten-fold increase in soil-derived particulates compared to museums without carpets and fewer visitors.

Particles do not necessarily adhere to walls when they touch. The four main adhesion mechanisms are capillary attraction (particularly for water-based and oil-based aerosols), electrostatic attraction (particularly for soot and lint), chemical reaction and physical interlocking. Chemical reaction and physical interlocking are initially (within the first fraction of a second) poor adhesion mechanisms, and large particles may easily bounce off due to momentum and elasticity, or be twisted off by gravity-induced torque. Rough surfaces are more effective in holding fine particles. Paints or surfaces (such as smooth glass) that resist chemical reaction and physical interlocking can reduce the total amount of material deposited.

In the case of ceilings or other downward-facing surfaces, gravity can remove particles when they dry out. It can also remove particles from vertical and sloping surfaces by applying a torque to the attachment, the top part of the attachment failing in tension.

6.5.2. From exterior surfaces

Experimental studies have indicated that surface cleaning of deposited salts by wind and condensation drip-off is of limited efficiency, and that surface cleaning by rain is the predominant mechanism (Cole *et al.*, 2004e). Surface cleaning occurs when raindrops run off the surface, collecting surface salts that are in their path. Run-off occurs when a raindrop on a surface has grown through coalescence to a critical size. When the slope of the surface is θ (0° for horizontal and 90° for vertical) and the maximum contact angle between the water and the surface is ϕ (the minimum contact angle is assumed to be zero here), then the maximum volume of a drop that can be retained on the surface V_{max} is calculated from (all dimensions are in mm):

$$A_{max} = 7.49 \, \frac{1-\cos(\phi)}{\sin(\theta)} \tag{37}$$

$$h_{max} = 1.2(1-(\cos(\phi))^{0.8}\left(1+\frac{0.5}{\sin(1.06\theta)}\right) \tag{38}$$

$$V_{max} = \frac{8}{3\pi}\frac{A_{max}^2}{h_{max}} \tag{39}$$

Figure 12 presents the volume of a raindrop required for run-off against the contact angle. The figure indicates that the more wettable a surface is (the lower the maximum contact

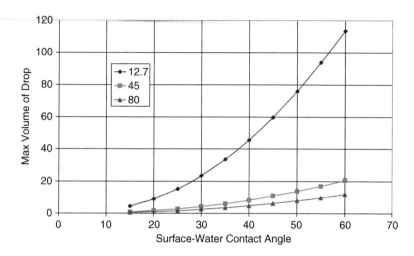

Fig. 12. Droplet run-off from a surface as a function of the surface–water contact angle ϕ for three surface slopes of $\theta = 12.7$, 45 and 80.

angle ϕ), the lower the drop volume before instability. Thus, more-wettable surfaces will clean better in lighter rain than less-wettable surfaces.

The results of these studies are most relevant to cultural objects exposed in the open. The results indicated that below a given amount of rain, the rain will be retained on the surface and may promote salt concentration rather than cleaning. If rain is in excess of this critical value, wash-off will occur and the surface will be cleaned. The volume of rain deposited in a "shower" shows a relationship to latitude, and thus the average volume of rain per shower is significantly lower in Hobart relative to Cairns. Thus, maintenance practices in Hobart should not rely on cleaning by rain, despite the relatively high rainfall is in this city. A second effect relates to the wettability of a surface. As indicated in Fig. 12, the higher the surface contact angle, the greater the maximum volume of a drop before run-off, and thus the greater the volume of rain before wash-off. Thus, surface treatments that reduce wettability will also reduce the effectiveness of rain washing.

6.6. Factors controlling gaseous pollutant levels within museums

A generalized interpretation of gaseous pollutant interactions within the museum environment is presented in Fig. 13. It describes how the indoor gaseous condition is comprised of contributions from the outdoor environment, but modified by additional contribution from the indoor environment materials, while at the same time absorbent materials indoors can reduce the total pollutant load. A contained environment such as a storage box or display case duplicates this response in the inner shell. Each level of containment provides buffering and dampening between the two environments, depending on the ventilation rate between the two.

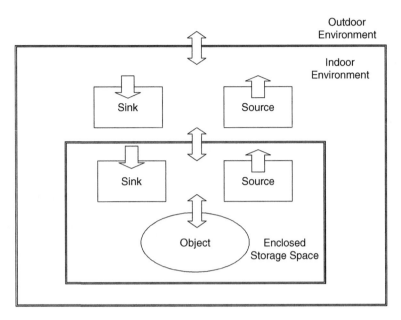

Fig. 13. Model of gaseous pollutant interactions within a museum.

The contribution from indoor sources is influenced by the dynamic pollutant emission behavior of materials Tichenor and Guo (1987) describe emission decay with the relationship:

$$R_t = k_1 M_0 \exp{(-k_1 t)} \tag{40}$$

where R_t is the emission rate at time t; k_1 is the first-order rate constant for emission decay and M_0 is the mass of pollutant that is present in the source at time $t = 0$ (note that $R_0 = k_1 M_0$). Table 5 details emission decay parameters for common surface coverings.

The overall concentration rate in an interior space is influenced by the rate of ventilation, volume of air within a cabinet or room space and the emission rate R_t and, as a result, most incidences of pollutant-induced corrosion in museum spaces occur in confined spaces such as display cases and cabinets.

The inside and outside concentration ratio is described by Weschler *et al.* (1989), with

$$\frac{C_I}{C_O} = \frac{N}{(KS/V + N)} \tag{41}$$

where C_I and C_O are the inside and outside concentrations ($\mu g/m^3$), N is the air exchange rate (l/h), K is the mass transfer coefficient of the pollutant/material system (also known as the deposition velocity; V_{dep}), S is the surface area of the material (m^2) and V is the net volume of air in the enclosed space (m^3).

Table 5. First order emission decay parameters for materials (from Brown, 2003)

Material	Pollutant	k_1 (h^{-1})	M_o (mg/m^2)
Acrylic paint A	Texanol®*	0.11	790
	TVOC	0.13	1800
Acrylic paint B	Texanol®	0.08	260
	TVOC	0.11	730
Low-odor acrylic paint	Propylene glycol	0.13	2400
	TVOC	0.11	730
Zero-VOC acrylic paint	2-Butoxyethanol	3.6	3.6
	TVOC	2.3	24
Enamel paint	Xylene isomers	1.6	1400
	TVOC	0.38	20000
"Natural" paint	Limonene	1.2	470
	TVOC	1.3	870
Wool carpet (new)	4-Phenylcyclohexene	0.003	28
	Styrene	0.068	18
	TVOC	0.051	58
Particle board (new)	Formaldehyde	0.001	380
	Methanol	0.006	520
	Hexanal	0.0002	260
	TVOC	0.014	84
MDF (new)	Formaldehyde	0.0006	650

* Texanol® is a proprietary coalescent aid for latex paint.

Primary sources for indoor gaseous pollutants are numerous, but can be generalized as building materials used for construction, including wood and wood products (especially particle and fiber board, which contains formaldehyde-based resins), adhesives, paints and coatings, polymer-containing furnishings and fabrics, cleaning products, anthropogenic sources and some objects in the collection. Printers and photocopy machines produce the majority of ozone in an indoor environment.

6.6.1. Implications to cultural practice of pollutant sources and deposition

Given the interdependence of gaseous pollutants, particulates and surfaces, no single factor can easily be considered in isolation. Surface wetting by micro droplet condensation nucleating around deliquescent salts promotes the formation of water, which will directly influence gaseous absorption into the droplet. Because of the equilibrium-driven dissolution of gases into droplets, the concentration of related species in a reaction series is favored, and can easily, often repeatedly, form a concentrated or saturated solution through dehydration induced by temperature or RH cycles. It has been seen repeatedly that these effects can result in the growth of surface efflorescence or "bloom," the enhancement of biological decay, corrosion and degradation of organic materials through hydrolysis and oxidation. The local environment at the surface is evidence of the nearby environment, itself influenced by the meso- and macro-scale environments that it is a subset of.

Strategies for dealing with pollutants must involve examining the effect of exposure (material dependent), along with additional contributing factors such as temperature, RH, UV and visible radiation and the presence of additional sources or sinks in the immediate vicinity. The consideration of all these factors in the context of a risk assessment, possibly on a range of scales, will need to be performed in conjunction with a monitoring program. Many control strategies exist: filtering, using sorbents, barriers, increasing ventilation rates, reducing exposure time, using enclosures, etc., all potentially subject to a cost–benefit analysis.

7. SURFACE FORMS AND DEGRADATION

7.1. Oxide products

Previously, we discussed the fact that aerosol size depended on its source (surf or white-caps), but as marine aerosols are hygroscopic, their size also depends on ambient RH. Further, when an aerosol first breaks free of a wave, it has a seawater composition, and then it gradually equilibrates (and thus decreases in size). Thus, marine aerosols may take four forms (Cole *et al.*, 2004b) – non-equilibrium, near-ocean aerosol (size range 6–300 μm), wet aerosol (3–150 μm), partially wet aerosol (1–60 μm) and dry aerosol (<1–20 μm) – depending on time of flight and ambient RH. These size ranges are based on aerosol mass or volume; mean sizes based on the number or the surface area of aerosols are much smaller. When these aerosols are deposited on a metal surface, a number of characteristic surface "forms" result from the surface–aerosol interaction (Cole *et al.*, 2004b). These forms differ in the extent of retained salts, degree of surface alteration and in the formation of corrosion nodules. For example, when a wet aerosol impacts on an aluminum surface (limited initial reactivity), a cluster of deposited salt crystals forms. These crystals have compositions of either NaCl, MgCl or $CaSO_4$, indicating that the original seawater solutions have segregated. In contrast, if the same aerosol impacts on a galvanized steel surface, there will be strong oxide growth on the surface (predominately simonkolleite and gordaite), with the retention of a NaCl crystal on this oxide layer. Interestingly, rather than the clean crystal edges that are observed for the NaCl crystal formation on aluminum, the NaCl crystal on galvanized steel appears to blend into the underlying oxide. Further oxide formation tends to be favored at the grain boundaries and triple points on the galvanized surface (see Fig. 14).

Recent work by Cole *et al.* (2004c) investigated the phases that form when microliter saline drops were placed on zinc. This study demonstrated the variety of corrosion products that may form, and highlighted the importance of mixed cation products in a situation where Na and Mg concentrations are several orders of magnitude higher than the Zn concentration generated by anodic activity. The study also highlighted that when dealing with microliter droplets, processes within the droplet (anodic and cathodic activity, mass transport and diffusion) can dramatically alter droplet chemistry and lead to corrosion products that would not be expected from the initial conditions.

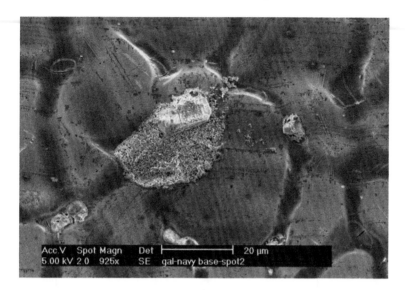

Fig. 14. SEM micrograph showing increased activity between salts and a galvanized steel surface at grain boundaries and triple points.

The implications of these observations to the conservation of metallic objects are both direct and indirect. The study, of course, provides direct evidence for mechanisms of zinc corrosion in marine environments. It also reinforces that when considering objects exposed to marine conditions, consideration must be given to the size range of aerosols and to the chemistry of marine aerosol. Misleading results can be obtained if marine exposures are approximated with immersion or NaCl-only exposure. The unique dynamics of droplets are relevant not only to marine locations, but to all cases where corrosion is promoted by localized wetting or deposition of rain aerosol or hygroscopic particulates. Droplets with volumes in submicroliters can undergo significant changes in chemistry, unlike corrosion in immersed situations, and the chemical changes can either enhance or restrict corrosion.

Further, the studies indicate that the effectiveness of maintenance strategies will vary significantly with reactivity of the metal components of the object. For example, a strategy of frequent washing to decrease salt may have little effect on a very reactive metal such as zinc, since the degradation occurs immediately after the deposition of a marine aerosol, but it may be quite effective for aluminum objects where a significant number of cycles of hygroscopic wetting of deposited salts is required to induce damage.

7.2. Implications of pollutants to object degradation

In Fig. 15, the ion concentration in a droplet exposed to conditions of CO_2 at 400 ppm, SO_2 from 20, 40, 75, 150 and 300 ppb and NH_3 at 20 ppb is given (Cole, 2000).

This examination of pollutant deposition and aqueous chemistry has a number of implications to the conservation of metallic objects. Clearly, in the case of metallic objects

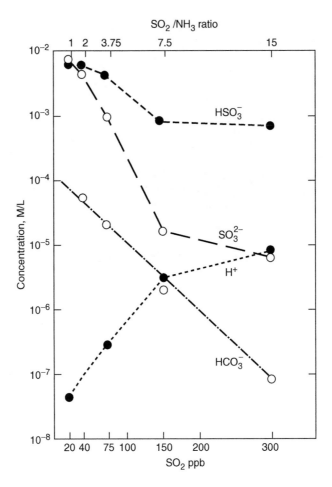

Fig. 15. Ion concentration as a function of gaseous SO_2 concentration and $SO_2:NH_3$ ratio (from Cole, 2000).

located in the open, the implications are direct. In this case, the major implication is that a knowledge of the gaseous SO_x may not give a reliable measure of corrosivity. Deposition rates, which will depend highly on both rain and RH, and oxidants and catalysts (such as O_3, H_2O_2 and Mn(II), Fe(III) and NO_2), and any alkali precursors (*e.g.* ammonia), will control the pH of the resulting moisture films or drops.

In an interior environment, the possible deposition pathways for pollutants will be highly dependent on RH and on any hygroscopic particulates or aerosols. If RH is low and in the absence hygroscopic species, only direct gaseous deposition is possible. However, if the RH exceeds the deliquescent RH of particulates in the air or on metal surfaces, then pollutant deposition will be enhanced by the absorption of gaseous species into the aqueous phases that form when the particulates wet. The deliquescent RH of some common salts are given in Table 6.

Table 6. Deliquescent relative humidity for salts found in common aerosols (at 20°C) (from Seinfeld and Pandis, 1997)

Salt	DRH (%)
Na_2SO_4	84.2
NH_4Cl	80.0
$(NH_4)_2SO_4$	79.9
$NaCl$	75.3
$NaNO_3$	74.3
$(NH_4)_3H(SO_4)_2$	69.0
NH_4NO_3	61.8
$NaHSO_4$	52.0
$(NH_4)HSO_4$	40.0
$MgCl$	35.0

An example where these factors may be in play is in the black spots on brass. It is observed that such black spots are favored by the presence of high RH and hygroscopic dust particles (Weichert *et al.*, 2004). Further, SO_2 is readily absorbed into the aqueous phase (as indicated by its high H_A). Thus, under these conditions, acidic sulfide- and sulfate-containing moisture phases are likely to form. The reaction of moist aerosols or surface droplets of such a composition with bronze could well lead to spotting, although the corrosion products likely to form would be chalcanthite ($CuSO_4 \cdot 5H_2O$), antlerite ($Cu_3SO_4(OH)_{4)}$) and brochantite ($Cu_4SO_4(OH)_6$), rather than covellite (CuS).

8. IMPLICATIONS FOR DESIGN AND MAINTENANCE STRATEGIES

The above analysis highlights that deposition mechanism and thus locations are highly dependent on particle size, so different strategies may be required, depending on the particle size.

Nevertheless, it is possible to make some general observations. It is apparent that soot can deposit rapidly due to electrostatic attraction. Thus, soot concentration should be minimized, as should electrostatically charged surfaces. Air filtration systems can be used as long as these do not lead to unacceptable effects (*e.g.* ozone generation). Internal generation of soot, such as in kitchens, should be minimized by isolating kitchens from museums. Antistatic coatings can be used to minimize electrostatic attraction of soft furnishings.

As indicated in the analysis, most large particles will settle via gravity, and the largest will settle near entry points. Double doors and other strategies will limit resuspension and entry of the largest particulates. In terms of whether to choose a carpet or solid floor, a carpet has more "filtering" and more electrostatic attraction, and so is more effective at removing aerosols from the air but, on the other hand, it can facilitate more resuspension.

Air plumes from heaters and open doors or windows can lead to impact deposition. Clearly, such effects can be minimized by double doors to limit airflow through doors and restrictions on opening windows in areas where objects are stored or displayed. Objects should not be placed in the vicinity of heaters, or heating and air-conditioning points should be multiple and distributed to minimize the airflow promoted by any given inlet or heat source.

This analysis has clearly demonstrated that fine and potentially damaging pollutants may deposit on objects placed on walls if there is significant air circulation. Systems that cause air stratification should be considered. Camuffo (1998) indicates that if air-conditioning inlets are placed just above ground level (at a hight sufficient to discourage resuspension of particulates deposited on floors), then the entry of cool air will supplant the warm air and enforce natural stratification. In contrast, inappropriately positioned and overzealous heating and air conditioning will promote air circulation.

9. CONCLUSIONS

The chapter has outlined the application of holistic modeling to pollutant deposition within and exterior to museums. In doing so, the chapter has demonstrated the advantages of linking pollutant sources' transport, deposition and effect in generating a full understanding of the degradation mechanisms, and the appropriate strategies to minimize degradation of cultural objects.

In particular, the chapter has highlighted

1. the role of turbulent diffusion in controlling deposition of aerosol on the exterior of buildings, and how this may lead to increased degradation at building edges;
2. the role of vortex shedding in promoting the deposition of medium-size aerosol in relatively slow-moving airstreams. The potential of this mechanism to promote soiling of cultural objects on walls is highlighted;
3. the role of momentum-dominated impact in promoting deposition in airstreams caused by cooking, open doors and windows, and the care required to ensure that this does not lead to degradation of culturally significant objects and ceilings.

REFERENCES

Albero, S., Giavarini, C., Santarelli, M.L., Vodret, A., 2004. CFD modeling for the conservation of the gilded vault hall in the domus aurea. *J. Cult. Heritage* **5**(2), 197–203.

Annis, P.J., 1991 Fine particle pollution; residential air quality. *North Central Regional Publication 393.*

Baer, N.S., Banks, P.N., 1985. Indoor air-pollution – effects on cultural and historic materials. *Int. J. Museum Management and Curatorship* **4**, 9–20.

Blackburn, H.M., Marques, F., Lopez, J.M., 2005. Symmetry breaking of two-dimensional time-periodic wakes. *J. Fluid Mech.* **522**, 395–411.

Brimblecombe, P., 1990. The composition of museum atmospheres. *Atmos. Environ.* **24B**(1), 1–8.

Brown, S.K., 2003. Indoor air pollution – lowering emissions of chemicals released from manufactured products. In *Hazmat 2003 Conference*, Fire Protection Association Australia, Stamford Grand, N. Ryde, NSW, Australia, April 2003, pp. 29–30.

Brown, S.K., Sim, M.R., Abramson, M.J., Gray, C.N., 1994. Concentrations of volatile organic compounds in indoor air – a review. *Indoor Air* **4**, 123–134.

Cadle, R.D., 1965. *Particle Size: Theory and Industrial Applications.* Reinhold Publishing: New York, pp. 147–150.

Camuffo, D., 1998. *Microclimate for Cultural Heritage.* Elsevier: Amsterdam, pp. 235–292.

Camuffo, D., Van Grieken, R., Busse, H.-J., Sturaro, G., Valentino, A., Bernardi, A., Blades, N., Shooter, D., Gysels, K., Wiesser, M., Kim, O., Ulrych, O., 2001. Environmental monitoring in four European museums. *Atmos. Environ.* **35**(Suppl. 1), S127–S140.

Clift, R., Grace, J.R., Weber, M.E., 1978. *Bubbles, Drops and Particles.* Academic Press: New York.

Cole, I.S., 2000. Mechanism of atmospheric corrosion in tropical environments. In *Marine Corrosion in Tropical Environments*, ASTM STP 1399. Eds. Dean, S.W., Hernandez-Duque Delgadillo, G., Bushman, J.B. American Society for Testing and Materials: Philadelphia, PA, pp. 33–47.

Cole, I.S., Ganther, W.D., 1996. A preliminary investigation into airborne salinity adjacent to and within the envelope of Australian houses. *Construction Building Mat.* **10**(3), 203–207.

Cole, I.S., Paterson, D.A., 2004. Holistic model for atmospheric corrosion: Part 5 – factors controlling deposition of salt aerosol on candles, plates and buildings. *Corrosion Eng. Sci. Technol.* **39**(2), 125–130.

Cole, I.S., Paterson, D.A., 2006. Mathematical models of the dependence of surface temperatures of exposed metal plates on environmental parameters. *Corrosion Eng. Sci. Technol.* **41**(1), 67–76.

Cole, I.S., Paterson, D.A., Ganther, W.D., 2003. Holistic model for atmospheric corrosion: Part 1 – theoretical framework for the production, transportation and deposition of marine salts. *Corrosion Eng. Sci. Technol.* **38**(2), 129–134.

Cole, I.S., Chan, W.Y., Trinidad, G.S., Paterson, D.A., 2004a. Holistic model for atmospheric corrosion: Part 4 – a geographic information system for predicting airborne salinity. *Corrosion Eng. Sci. Technol.* **39**(1), 89–96.

Cole, I.S., Lau, D., Paterson, D.A., 2004b. Holistic model for atmospheric corrosion: Part 6 – from wet aerosol to salt deposit. *Corrosion Eng. Sci. Technol.* **39**(3), 209–218

Cole, I.S., Muster, T.H., Paterson, D.A., Furman, S.A., Trinidad, G.S., Wright, N., 2004c. Holistic model of atmospheric corrosion: extending a microclimatic model into a true corrosion model. In *Eurocorr 2004 Conference*, Nice, France, 12–16 September.

Cole, I.S., Ganther, W.D, Sinclair, J.D., Lau, D., Paterson, D.A., 2004d. A study of the wetting of metal surfaces in order to understand the processes controlling atmospheric corrosion. *J. Electrochem. Soc.* **151**(12), B627–B635.

Cole, I.S., Lau, D., Chan, F., Paterson, D.A., 2004e. Experimental studies of salts removal from metal surfaces by wind and rain. *Corrosion Eng. Sci. Technol.* **39**(3), 333–338.

Cole, I.S., Ganther, W.D., Paterson, D.A., Bradbury, A., 2005a. Experimental studies on dependence of surface temperatures of exposed metal plates on environmental parameters. *Corrosion Eng. Sci. Technol.* **40**(4), 328–336.

Corvo, F., Betancourt, N., Mendoza, A., 1995. The influence of airborne salinity on the atmospheric corrosion of steel. *Corrosion Sci.* **37**, 1889–1901.

Corvo, F., Hayes, C., Betancourt, N., Maldonado, L., Veleva, L., Echeverria, M., de Rincon, O., Rincon, A., 1997. *Corrosion Sci.* **39**, 823–833.

De Santis, F., Di Paolo, V., Allegrini, I., 1992. Determination of some atmospheric pollutants inside a museum: relationship with the concentration outside. *Sci. Total Env.* **127**, 211–223.

Drakou, G., Zerefos, C., Ziomas, I., Voyatzaki, M. 1998. Measurements and numerical simulations of indoor O_3 and NO_x in two different cases. *Atmos. Environ.* **32**(4), 595–610.

European Organisation for Technical Approvals (EOTA), 1997. *Assessment of Working Life of Products: Part 3 – Durability*, TB97/24/9.3.1. EOTA: Brussels, Belgium.

Goldman, M., 1978. The radiometer revisited. *Phys. Educ.* **13**, 427–429.

Gustafsson, M.E.R., Franzen, L.G., (1996). Dry deposition and concentration of marine aerosols in a coastal area, SW Sweden. *Atmos. Environ.* **30**, 977–989.

Gysels, K., Delalieux, F., Deutsch, F., Van Grieken, R., Camuffo, D., Bernardi, A., Sturaro, G., Busse, H.J., Wieser, M., 2004. Indoor environment and conservation in the Royal Museum of Fine Arts, Antwerp, Belgium. *J. Cult. Heritage* **5**(2), 221–230.

Hill, L.J.K., Bouwmeester, W., 1994. *Scottish Museums Council Factsheet – Air Pollution.* Scottish Museums Council: Edinburgh.

Hoffman, M.R., Calvert, J.G., 1985. *Chemical Transformation Modules for Eulerian Acid Deposition Models: Vol. 2 – The Aqueous-Phase Chemistry*, EPA/600/3-85/017. US Environmental Protection Agency: Research Triangle Park NC.

Injuk, J., Osan, J., Van Greiken, R., Tsuji, K., 2002. Airborne particles in the Miyagi Museum of Art in Sendai, Japan, studied by electron probe X-ray microanalysis and energy dispersive X-ray fluorescence analysis. *Anal. Sci.* **18**(5), 561–566.

Johnson, K.E., Stanners, J.F., 1981. The characteristics of corrosion test sites in the community. *Report EUR 7433.* Commission of The European Communities: Luxembourg.

Katsanos, N.A., De Santis, F., Cordoba. A., Roubani-Kalantzopoulou. F., Pasella, D., 1999. Corrosive effects from the deposition of gaseous pollutants on surfaces of cultural and artistic value inside museums. *J. Hazard. Mater. A* **64**, 21–36.

Laitone, J.A., 1979. Erosion prediction near a stagnation point resulting from aerodynamically entrained solid particles. *J. Aircr.* **16**, 809–814.

Morcillo, M., Chico, B., Otero, E., 1999. Effect of marine aerosol on atmospheric corrosion. *Mater. Perform.* **38**(4), 72–77.

Morcillo, M., Chico, B., Mariaca, L., Otero, E., 2000. Salinity in marine atmospheric corrosion: its dependence on the wind regime existing in the site. *Corrosion Sci.* **42**(1), 91–104.

Ohba, R., Okabayashi, K., Yamamoto, M., Tsuru, M., 1990. A method for predicting the content of sea salt particles in the atmosphere. *Atmos. Environ. A.* **24A**, 925–935.

Papakonstantinou, K.A., Kiranoudis, C.T., Markatos, N.C., 2000. Mathematical modeling of environmental conditions inside historical buildings: the case of the archaeological museum of Athens. *Energy Buildings* **31**(3), 211–220.

Prado, G.P., Howard, J.B., 1978. Formation of large hydrocarbon ions in sooting flames, evaporation-combustion of fuels. *Adv. Chem. Ser.* **166**, 153–166.

Rhyl-Svendsen, M., Glastrup, J., 2002. Acetic acid and formic acid concentrations in the museum environment by SPME-GC/MS. *Atmos. Environ.* **36**, 3909–3916.

Salmon, L.G., Cass, G.R., Bruckman, K., Haber, J., 2000. Ozone exposure inside museums in the historic central district of Krakow, Poland. *Atmos. Environ.* **34**(22), 3823–3832.

Seinfeld, J., Pandis, S., 1997. *Atmospheric Chemistry and Physics: From Air Pollution to Climate Change.* Wiley Interscience, New York.

Sheard, G.J., Thompson, M.C., Hourigan, K., Leweke, T., 2005. The evolution of a subharmonic mode in a vortex street. *J. Fluid Mech.* **534**, 23–38.

Sohankar, A., Davidson, L., Norberg, C., 1995. Vortex shedding: numerical simulation of unsteady flow around a square two-dimensional cylinder. In *12th Australian Fluid Mechanics Conference*, Sydney, Australia, pp. 517–520.

Strekalov, P.V., Panchenko, Y., 1994. The role of marine aerosols in atmospheric corrosion of metals. *Prot. Met.* **30**, 254.

Thatcher, T.L., Laia, A.C.K., Moreno-Jacksona, R., Sextro, R.G., Nazaroffa, W.W., 2002. Effects of room furnishings and air speed on particle deposition rates indoors. *Atmos. Environ.* **36**(11), 1811–1819.

Tichenor, B.A., Guo, Z., 1987. The effect of ventilation rates of wood finishing materials. *Proc. Indoor Air* **87**(3), 423–432

Tuma, J.J., 1976. *Handbook of Physical Calculations.* McGraw Hill, New York.

Weichert, M., Eggert, G., Jones, M., Ankersmit, B., 2004. Trees, bunches, cauliflower – a closer look at sulphurous corrosion on copper alloys ("black spots"). *Metals*.

Weinkauf, T., Hege, H.-C., Noack, B.R., Schlegel, M., Dillmann, A., 2003. Coherent structures in a transitional flow around a backward-facing step. *Phys. Fluids* **15**(9), S3.

Weschler, C.J., Shields, H.C., 1999. Indoor ozone/terpene reactions as a source of indoor particles. *Atmos. Environ.* **33**(15), 2301–2312

Weschler, C.J., Shields, H.C., Naik, D.V., 1989. Indoor ozone exposures. *J. Air Pollut. Control Assoc.* **39**, 1562–1568.

Yoon, H.Y., Brimblecombe, P., 2001. The distribution of soiling by coarse particulate matter in the museum environment. *Indoor Air* **11**, 232–240.

Chapter 4

Examples of Using Advanced Analytical Techniques to Investigate the Degradation of Photographic Materials

Giovanna Di Pietro

Institute of Monument Conservation and Building Research, Swiss Federal Institute of Technology, Zürich, Switzerland
Email: dipietro@arch.ethz.ch

Abstract

This chapter reviews advanced analytical techniques developed and adapted for the study of photographic materials. Two examples of investigations into the degradation of photographic materials are given. Both make use of a wide range of spectroscopic techniques. Both investigations were aimed at understanding the fundamental mechanisms of degradation of photographic materials and were performed in different time periods and in different locations.

The first is a study of the phenomenon called silver mirroring, a common degradation of black-and-white photographs. The second is a study of the fading of motion picture films in the collection of the National Film and Sound Archive of Australia.

Keywords: Photographic emulsions, photographic dyes, Raman microscopy, FTIR, XPS, XRD, XRF, TEM.

Contents

Physical Techniques in the Study of Art, Archaeology and Cultural Heritage
Edited by D. Creagh and D. Bradley

1. INTRODUCTION

Photographic materials are a relevant part of our cultural heritage. Both still photographs and motion picture films are not only objects of art but also substantial tools for the reconstruction and understanding of our history. Their numbers in archives and museums are impressive but the number of photographic conservators and photographic conservation scientists is relatively small. Despite this, the community of photographic conservators and conservation scientists worldwide has produced high-quality research in the last 20 years. This research has helped us to take a significant step forward in understanding the degradation mechanisms of photographic materials.

A detailed understanding of the mechanisms of deterioration of the wide variety of photographic materials is essential to set a rational basis for preventive conservation measures for photographs. This detailed understanding is gained firstly through scientific examination of photographic materials and consequently through scientific studies of their degradation.

Photographic materials are layered structures composed for the most part of a support and of an image layer. The support can be transparent (as in motion picture films and in negatives in still photography) or opaque (like in prints in still photography). The support can be made of glass, synthetic polymer, or paper. The most common transparent supports used today are made of cellulose triacetate and polyester, whereas the most common opaque supports are made of baryta paper (which is a paper coated with barium sulphate to obtain a smooth, high-reflectance finish) and resin-coated paper (which is a photographic paper that has been coated with polyethylene on both sides). The image layer is a gelatin emulsion. In the case of black-and-white images, the emulsion contains silver halide crystals, which are turned to silver grains during processing. The silver grains have dimensions on the order of microns and, in chemically developed photographs, a complex dendritic form. The small dimension and the presence of kinks result in their having a very large surface area. This explains their susceptibility to undergo reactions and corrosion processes. In the case of colour images, the emulsion is made of three different layers, each of them containing either cyan, magenta, or yellow dyes, and together constituting the final image. The fading characteristics of these dyes depend heavily on their precise chemical structure, and on the presence of layers filtering out specific regions of the visible spectrum of light. In Fig. 1, the cross section of a typical black-and-white negative is presented, and in Fig. 2, the cross section of a typical colour print in motion picture films is shown.

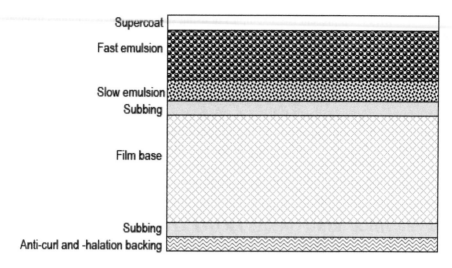

Fig. 1. Structure of a typical black-and-white photographic negative.

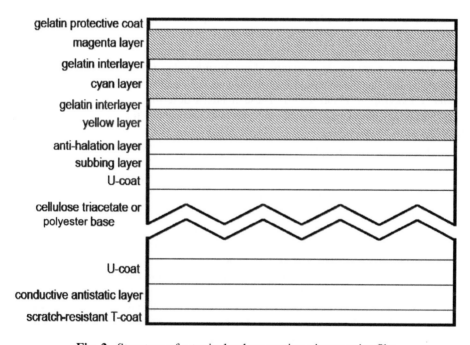

Fig. 2. Structure of a typical colour motion picture print film.

In this chapter, I shall present two examples of using spectroscopic techniques to investigate the degradation of photographic materials. Both investigations were aimed at understanding the fundamental mechanisms of degradation of photographic materials and were performed in different time periods and in different locations.

My first example is a study of the phenomenon called silver mirroring, a common degradation of black-and-white photographs. I performed this study as part of my PhD research at the Laboratory for Image and Media Technology of the University of Basel in the years 1998–2002 (Di Pietro and Ligterink 2002; Di Pietro, 2004).

The second example is a study of the fading of motion picture films. I undertook this research in 2004 at the University of Canberra in collaboration with the National Film and Sound Archive in Australia (Di Pietro *et al.*, 2005).

2. SILVER MIRRORING ON SILVER GELATIN GLASS PLATES

Silver mirroring is a bluish metallic sheen appearing on the surface of silver-based photographs as a result of ageing. One of the photographic processes most affected by silver mirroring is that of silver gelatin glass negatives, the most common photographic negative process between the 1880s and the 1920s.

Silver gelatin glass negatives have as support a glass plate of thickness 1–2 mm. The emulsion, whose thickness is on the order of 50 μm, is applied on one side of the plate, which is made of gelatin and, in most cases, of silver bromide grains, turn to metallic silver after chemical development. Two valuable historical sources on the manufacture of silver gelatin glass plates are the third volume of the *Ausführliches Handbuch der Photographie* (Eder, 1903) and the fourth volume of the *Handbuch der wissenschaftlichen und angewandten Photographie* (Jahr, 1930). These manuals describe how to choose, cut, and prepare the glass plates and how to mix, prepare, and apply the emulsion layer, both manually and with machines.

A review on the conservation-related problems of silver gelatin glass plates can be found in the article of Gillet *et al.* (1986).

Silver mirroring is usually present at the edges of the plate. This degradation pattern is so common that it is sometimes used to distinguish this photographic process from other types of glass negatives. Figures 3 and 4 present two examples of silver mirroring on silver gelatin glass negatives. Silver mirroring is sometimes present at the centre of the plate with patterns resembling the surface characteristics of the envelopes in which they were stored (Fig. 5).

Both the presence of silver mirroring at the edges of the plates and the connection between silver mirroring and the surface characteristics of the envelope suggest that the formation of silver mirroring is closely connected to how the plates are stored. The aim of this work is to arrive at a better understanding of the mechanisms of formation of silver mirroring in order to choose the most suitable enclosure materials and storage conditions on a more rational basis than has been used hitherto.

In the following paragraphs, I will first present the models developed in the last 150 years for the formation of silver mirroring, and then all the analyses that have been performed to answer the questions still open on its mechanism of formation. Based on the results of these analyses, a modification to the current model of silver mirroring formation is presented.

Fig. 3. Silver gelatin glass negative. Cueni study collection (~1910). The mirroring sheen is narrow and blunt, just visible at the plate edges.

2.1. Historical review of the models on silver mirroring formation

The first account of silver mirroring dates back to 1882 (British Journal of Photography, 1982), just two years after the invention of modern black-and-white photography. In these early stages, silver mirroring was attributed to the action of hydrogen sulphide ("sulphuretted hydrogen") (British Journal of Photography, 1901), and the formation of silver mirroring was defined as a "slow sulphiding" (British Journal of Photography, 1918).

In an article in the British Journal of Photography (1922), a second compound is recognised as playing an important role in the formation of silver mirroring, *i.e.* silver salts left in the emulsion by incomplete fixing. Silver mirroring would result from the reaction between sulphur-containing compounds and these residual silver salts.

The importance of incomplete fixing is the core idea behind the articles published by Shaw in 1931 (Shaw, 1931a,b), where silver mirroring is called "tarnishing", in accordance

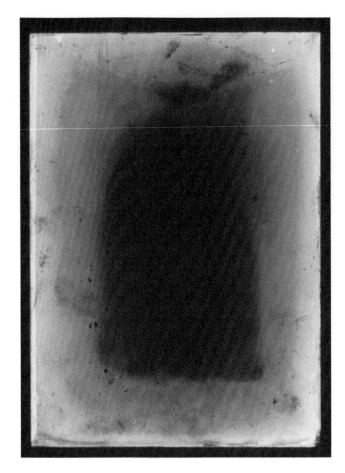

Fig. 4. Silver gelatin glass negative. Cueni study collection (~1910). The mirroring stain is wide, partially obscuring the image.

with the term used for the degradation of silver plates and of daguerreotypes. In these articles, the role played by hydrogen sulphide is denied. Silver mirroring would result from an unspecified attack of the image by the products of decomposition of the "silver–sodium–halide–thiosulphate complexes", the decomposition being triggered by atmospheric CO_2.

The research on the nature and mechanism of silver mirroring formation improved in the 1960s due to the observation that silver mirroring was often associated to what was felt as a great danger for our heritage, *i.e.* the appearance of red spots on microfilms (Henn and Wiest, 1963; McCamy and Pope, 1965; McCamy *et al.*, 1969). In 1963, Henn and Wiest proposed a model called in the present work as the *oxidation–migration–re-aggregation model*. First, the image silver particles are oxidised and the resulting silver ions migrate into the gelatin, then the silver ions are reduced to silver and they re-aggregate either

Fig. 5. Silver gelatin glass negative (courtesy of S. Dobrusskin). The regular three-time repetitive spot pattern suggests a specific process of formation: the envelope has some local feature causing silver mirroring and, as it is slightly bigger than the plate, it creates mirroring spots at slightly different positions every time the plate is displaced. Along the upper side, the shape of the mirroring stain matches the rounded opening of a sleeve.

within the gelatin, forming small particles appearing as red spots, or at the top surface of the emulsion forming silver mirroring (Henn and Wiest, 1963). The main difference with the previous ideas is that silver ions are not believed to be due to residues resulting from careless processing but to reaction with external agents such as oxidising gases. Peroxides or atmospheric oxygen in combination with hydrogen sulphide, ammonia, and sulphur dioxide were considered to be both oxidant and reducing compounds. Henn and Wiest showed that the red spots could be artificially created by exposing microfilms to vapours of hydrogen peroxide. They investigated making use of electron microprobe X-ray analysis for finding the composition of mirrored areas, concluding that they contained "appreciable but highly variable quantities of silver sulphide".

The three steps of oxidation, migration, and re-aggregation constitute the basic framework for almost all the research on silver mirroring formation published later. In 1981, Feldman, in a work on the discoloration of photographic prints, proposed some changes to the oxidation–migration–re-aggregation model as expressed by Henn and Wiest. Silver ions, as a result of the reaction with peroxides, could either be reduced to metallic silver by light, or could react with hydrogen sulphide to give silver sulphide (Feldman, 1981). He published the first transmission electron micrographs of mirrored photographs showing that silver mirroring consists of a layer of colloidal particles clustered at the top surface of the emulsion.

A researcher very active in this field was Klaus B. Hendriks of the Public Archives of Canada. He supported the oxidation–migration–re-aggregation model with transmission electron micrographs (Hendriks, 1989, 1991a,b). First, he showed that the image silver grains in oxidised silver gelatin emulsions are often surrounded by clouds of smaller particles. This confirmed the theory of Torigoe and co-workers that the consequence of image oxidation is the formation of smaller colloidal particles causing yellow discoloration to the photograph (Torigoe *et al.*, 1984).

Second, he confirmed the results of Feldman, showing that mirrored areas are characterised by colloidal particles at the top surface of the emulsion. Hendriks believed that these particles were proof of an upward migration of the silver ions. He added that the presence of transfer images in the baryta layer of deteriorated silver gelatin prints, already noticed by Weyde in the 1950s (Weyde, 1955), was proof of a downward migration of the silver ions.[1] He thought the silver mirroring particles to be made of a very thin layer of elementary silver, centred on a silver sulphide nucleus (Hendriks, 1984). He indicated the following as compounds responsible for silver mirroring:

(a) Compounds present either in the image layer or in the support as a result of careless processing;
(b) Atmospheric gases such as sulphur dioxide, hydrogen sulphide, oxides of nitrogen, peroxides, and ozone; and
(c) Compounds present in the filing enclosures in contact with the photograph.

Images published by Hendriks provide strong evidence that the first step to image deterioration is the oxidation of the image grains. Image grains lose their integrity, and smaller particles are formed within the emulsion or at the emulsion surface. Nevertheless, Hendriks did not explain why the ions or the colloidal particles would migrate towards the emulsion surface. And he did not explain why, in some cases, small particles are formed within the emulsion. These result in macroscopic yellowing discoloration, and in other cases the particles are formed at the top surface of the emulsion, resulting in silver mirroring.

An attempt to answer these crucial questions and the question of the chemical nature of silver mirroring was made by Nielsen and Lavedrine (Nielsen, 1993; Nielsen and Lavedrine, 1993). They published transmission electron micrographs of historically and artificially mirrored photographs showing that surface particles were present only in the mirrored regions and that smaller particles are found underneath the top layer of closely packed mirroring particles. The concentration and size of the smaller particles decrease as the distance from the surface increases. This particle distribution is considered as a proof for the migration of silver salts towards the surface, although no driving force for such a migration is given. The close-fitting mosaic of the uppermost particles is indicated as a proof of their gradual growth from smaller nuclei. They did not define the chemical composition of silver mirroring because only qualitative scanning electron microscope-energy

---

[1] "Transfer images" (also called "ghost images") are silver-based images formed spontaneously, during processing, on the baryta layer of prints processed with developing solutions contaminated with fixing solutions. They were reported for the first time by Weyde (1955). They are formed in wet emulsions when the fixer dissolves the not-exposed silver halide grains and the resulting silver ions are reduced back to silver by the developer. The reduction of the silver ions by the developer is normally not possible in solution but it can take place when the ions are attached to particles present in the baryta layer. This mechanism is at the base of diffusion transfer photographic processes, as in Polaroid photographs.

dispersive X-rays (SEM-EDX) analysis was performed. The analysis detected the presence of both silver and sulphur.

Finally, it is important to discuss an article written in 1988 that proposes a theory of silver mirroring formation drastically different from the ideas developed in the twentieth century. The theory presented by Barger and Hill (1988) is based on the fact that both surface roughness measurements and scanning electron micrographs showed that the mirrored areas have a higher surface roughness in comparison with non-mirrored areas. Moreover, their SEM-EDX analysis detected, in the mirrored areas, the presence of silver. These results, combined with the relevant difference between the reflectance spectrum of mirrored emulsions and of silver sulphide or of silver films, prompted them to propose a non-chemical mechanism of silver mirroring formation. The bluish appearance of mirrored photographs would result from a shrinking of the gelatin around the image particles. Such a rough surface would scatter the light differently and, therefore, would acquire a bluish tone.

2.2. Open questions on silver mirroring

The established model for silver mirroring formation among the community of photographic conservators is the oxidation–migration–re-aggregation model. The work of Nielsen and Lavedrine has clearly demonstrated that particles are present at the top surface of the emulsion only in the visually mirrored areas. This is in contradiction with the gelatin shrinking model of Burger and Hill.

Nevertheless, the oxidation–migration–re-aggregation model leaves two main questions to be answered:

1. What is the chemical composition of silver mirroring and, therefore, which are the compounds responsible apart from oxidant compounds?
2. Which is the driving force for the formation and/or aggregation of particles at the emulsion surface and, therefore, why is there the formation of silver mirroring in certain cases and yellowing discoloration in others?

Answering the first question is crucial to understanding which compounds are actually responsible for the formation of silver mirroring. If the silver mirroring particles are made of elementary silver (Ag), then the compounds responsible for silver mirroring, apart from oxidant compounds, have to be searched for among silver-reducing substances (*e.g.* aldehydes), while if they are made of silver sulphide (Ag_2S), they have to be searched for among sulphur containing substances (*e.g.* hydrogen sulphide (H_2S)). This question will be answered by analysing the chemical composition of the silver mirroring layer with spectroscopic techniques.

Answering the second question is crucial to distinguishing the formation of silver mirroring, which has been shown microscopically to be a layer of colloidal particles clustered at the top surface of the emulsion, from the other corrosion-based degradation forms of black-and-white photographs, which microscopically consist of small particles aggregated within the emulsion bulk. This question will be answered by analysing the particle shape and the density of the particle distribution underneath the mirroring layer. The particle shape and density are properties inherited by the mechanism of formation of silver mirroring.

A surface layer of particles, like silver mirroring, can result either from particles formed within the emulsion bulk and then migrating towards the emulsion surface or from particles formed directly at the emulsion surface. These two mechanisms lead to two different trends for the particle shape.

In the first case, the particle shape is predicted to be spherical at great distances from the surface and ellipsoidal close to the surface, as colloidal particles cannot escape from the emulsion and cluster once their spatial density is sufficiently high.

In the second case, particles would be formed by the reaction between silver ions, results of the oxidation of the image grains, and an external compound, and, as the reaction has no preferred direction, their shape is predicted to be spherical at any distance from the surface.

2.3. Experiments

The following experiments were performed on silver gelatin glass negatives belonging to the Cueni study collection, a collection of about 150 glass plate negatives of the Swiss painter and amateur photographer Adolf Cueni active in the Basel region in the years 1910–1920s. The Laboratory of Image and Media Technology of the University of Basel presently owns the plates.

2.3.1. Choice of spectroscopic techniques for the analysis of the chemical composition of silver mirroring

The spectroscopic methods suited to determine the chemical composition of silver mirroring have to fulfil two prerequisites.

The first prerequisite is that they have to distinguish between compounds present in the mirroring layer, of thickness on the order of hundreds of nanometres, and compounds present in the emulsion bulk, of thickness of the order of 50 μm.

The second prerequisite is that they have to be able to detect silver sulphide. The simple detection of sulphur is not sufficient to draw valid conclusions. Indeed, sulphur compounds can arise from the protein cysteine, one of the constituent proteins of the gelatin, and from the incomplete removal of sodium thiosulphate ($Na_2S_2O_3$), the usual fixing agent. In order to arrive at a conclusion, it is fundamental either to determine the amount of sulphur and silver in the mirroring layer (if the particles are made of silver sulphide, their amount has to be in the ratio 2:1) or to detect the presence of silver sulphide directly (e.g. detection of silver sulphide crystal structures).

Spectroscopic analyses are based on the detection of the radiation emitted by a sample excited by an incoming beam. The emitted radiation is characteristic of the elements and, in some cases, of the compounds present in the sample.

In spectroscopic equipments, samples of mirrored emulsions can be analysed using one of the following arrangements:

1. emulsion layer perpendicular to the incoming radiation (flat samples);
2. emulsion layer parallel to the incoming radiation (cross section samples); and
3. powder of emulsion scratched off from the mirroring layer (powder samples).

The fulfilment of the first prerequisite depends on both the properties of the incoming radiation beam and on the arrangement used.

Flat and cross section sample analyses give reliable results when the penetration depth and area of the incoming beam are smaller than the silver mirroring thickness, respectively (Fig. 6). Powder sample analyses are in every case reliable but require highly sensitive spectroscopic techniques due to the small amount of sample analysed.

It is therefore necessary to evaluate initially the penetration depth of the radiation used by the X-ray techniques available to the author, X-rays and electrons, into photographic emulsions.

When X-rays penetrate into a material, they are absorbed by different mechanisms (scattering, photoelectric effect, and pair production), so that the intensity I (−) of the incoming beam decreases with the distance z (cm) travelled in the material according to a simple exponential law:

$$I(z) = I_0 e^{-\mu \rho z} \tag{1}$$

where μ (cm^2 g^{-1}) is the mass attenuation coefficient and ρ (g cm^{-3}) is the material density. The penetration depth pd (cm) is defined as the distance at which the incoming beam has reduced its intensity to e^{-1} of the initial value:

$$\mathrm{pd} = \frac{1}{\mu \times \rho} \tag{2}$$

The mass attenuation coefficient depends on the energy of the incoming beam, and it is tabulated for most of the materials (Weast, 1985).

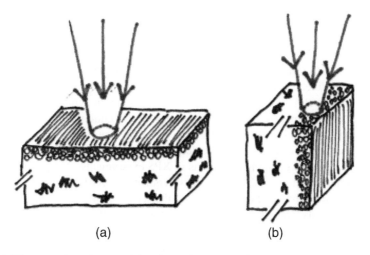

(a)   (b)

Fig. 6. (a) Flat samples: the emulsion layer is perpendicular to the incoming radiation. (b) Cross-sectional samples: the emulsion layer is parallel to the incoming radiation.

If the sample is composed of two materials A and B present in volume fractions $\varphi(A)$ and $\varphi(B)$, the penetration depth will be:

$$\text{pd} = \frac{1}{\mu(A) \times \rho(A) \times \varphi(A) + \mu(B) \times \rho(B) \times \varphi(B)} \tag{3}$$

Assuming that an emulsion is made of 10% (by volume) of silver ($\mu(Ag) = 218$ cm^2 g^{-1}, $\rho(Ag) = 10.5$ g cm^{-3}) and 90% of gelatin ($\mu(gel) = 9.8$ cm^2 g^{-1} as for water, $\rho(gel) = 1.29$ g cm^{-3}), the penetration depth calculated for an incoming X-ray beam of 8.041 keV (energy of the most common X-ray source in X-ray diffraction (XRD) instruments) is of the order of 1 mm. The penetration depth in a mirrored emulsion will be smaller because the emulsion is covered by a layer of particles. Nevertheless, even the penetration depth in pure silver is on the order of 4 μm, about 20 times bigger than the thickness of the silver mirroring. Therefore, the flat sample arrangement is not suitable for XRD experiments.

On the other hand, X-ray equipment capable of focussing the beams are often not available in research laboratories, especially systems that can focus the X-ray beam to a diameter smaller than few microns. See Creagh (Chapter 1) for information on focussed beams in microspectroscopy at synchrotron radiation sources. In addition, the cross section arrangement of the specimen is not suited to this approach.

When electrons penetrate into a material, they lose their energy with different processes dependent on their energy value. If their kinetic energies are in the range of kilo electron volts, the main mechanism is ionisation or excitation of the atoms present in the material. The amount of energy loss per unit of travelled path is proportional to the electron's density in the material, such that materials made of heavy elements will stop electrons much faster than light element materials. The electron's penetration depth is estimated calculating the path travelled by the electrons before they stop. This is called the continuous slowing down approximation (CSDA) range, and it is tabulated for most materials.

For starting electron energies of 20 keV, the typical energy of the incoming electrons in an SEM X-ray microanalysis apparatus, the penetration depth is about 6 μm in photographic gelatin and 1.5 μm in silver, in both cases more than the thickness of the silver mirroring layer. Therefore, the flat sample arrangement is not suited for SEM experiments.

If the electron kinetic energy is on the order of a few hundreds of electron volts, the CSDA range is not valid. Few hundreds of electron volts are the typical energy of the electrons emitted in X-ray photoelectron spectroscopy (XPS) equipments. This energy is so small that uniquely the electrons emitted from atoms at a maximum distance of 10 nm from the sample surface can escape and be detected (Grunthaner, 1987). Therefore, the flat sample arrangement can give reliable results in XPS experiments.

In conclusion, the methods fulfilling the first requisite are XRD on powder samples and XPS on flat samples. Both methods also satisfy the second requisite because XRD detects crystalline compounds (and, therefore, directly the presence of silver sulphide), while XPS provides quantitative results of the sample atomic composition.

The combination of these X-ray analyses has been used to determine the chemical composition of the edge silver mirroring present on four glass negatives belonging to the Cueni study collection. Three out of the four plates examined (numbers 1, 2, and 4) were processed negatives, while the number 3 was a historical non-processed glass negative.

Table 1. Spectroscopic analyses

Samples	Analysis
Plate 1	
1a	XRD powder
Plate 2	
2a	XRD powder
2b	XRD flat
2c	XPS flat
2d	SEM-EDX
Plate 3	
3a	XRD flat
3b	XPS flat
Plate 4	
4a	SEM-EDX

Small, approximately squared samples were cut out from the plates with a diamond knife, cutting from the glass side. Every sample underwent only one type of analysis.

Table 1 summarises the experiments performed.

2.3.1.1. X-ray diffraction (XRD)

Principles of the technique.   The XRD analysis detects the crystal structures present in a sample. Every crystal in the sample has a set of characteristic distances between the planes on which the atoms are located. When an X-ray beam of wavelength λ hits the atoms, the rays reflected from the atoms located on two parallel planes combine additively if the distance d between the two planes satisfies the so-called Bragg's law:

$$n\lambda = 2d \sin(\theta) \tag{4}$$

where n is an integer number and θ is the angle between the incident (or the reflected) beam and the perpendicular to the plane.

By changing the angle θ, the distances between the planes are scanned. Peaks of intensity of the diffracted radiation are present in correspondence of the distances satisfying the relation (4).

In the case where the diffracted radiation is detected with a scintillation detector, the spectrum is a graph of the intensity of the diffracted X-ray towards the angle θ.

In the case where the diffracted radiation is recorded on a photographic film, the spectrum consists of circles located around the direction of the incoming beam of radii simply related to the characteristic distances.

Measurements on powder samples.   The XRD analysis on powder samples was performed at the XRD facilities of the Netherlands Institute for Cultural Heritage, Amsterdam, in collaboration with Mr. P. Hallebeek. The XRD instrument is composed of an X-ray generator (Philips PW 1010) using a Cukα X-ray source at wavelength $\lambda = 1.5406$ Å and energy $E = 8.041$ keV, using a Debey–Scherrer powder camera for which the recording medium was double-coated CEA Reflex 25 film. The sample, consisting of few powder grains, is fixed on the tip of a glass fibre with cedar oil and it is continuously rotated during the measuring time (order of a few hours) to cover all the possible mutual positions between the incoming beam and the crystal planes within the sample. This instrument is able to detect the crystal composition of extremely minute amount of sample, of dimension of the order of 0.5 mm^2.

The silver mirroring was scratched off the plates with a scalpel under a loupe, with careful attention paid to removing only the mirroring layer and not the underlying emulsion. The total amount of material was of the order of few powder grains.

Measurements on flat samples.   The XRD analysis on flat samples was performed at the XRD facility of the Institute of Inorganic Chemistry of the University of Zürich, in collaboration with Prof. H. Berke. The XRD diffractometer is a Kristalloflex instrument produced by Siemens using a Cukα X-ray source at wavelength $\lambda = 1.5406$ Å and energy $E = 8.041$ keV. The diffracted rays are collected with a scintillation detector. The relative position of the sample and the detector was changed in steps of $2\theta = 0.05°$ using a measuring time of 0.3 s for every step.

The samples have approximate dimensions of 2 cm $\times$ 1 cm and they are hold flat in aluminium holders. In order to prevent contributions in the diffracted beam from the side areas not presenting silver mirroring, these areas were covered with self-adhesive tape.

2.3.1.2. X-ray photoelectron spectroscopy (XPS)

Principles of the technique.   XPS is a spectroscopic technique based on the measurement of the binding energies (E_b) of the core electrons of the atoms contained in a material. It is able to detect all the atoms of the periodic table other than hydrogen.

The sample is bombarded with monochromatic X-rays of energy E_X higher than the core-electron-binding energies; the atoms present in the sample absorb the X-rays and eject their core electrons with a kinetic energy E_k satisfying the relation:

$$E_k = E_x - E_b \tag{5}$$

Since E_x is known and E_k is measured in the experiment, the core-electron-binding energies E_b are calculated and used to identify the atoms.

Shifts in the measured core-electron-binding energy can be used to determine the molecular composition of the sample. XPS is a real surface analysis because only electrons ejected from a maximum distance of 10 nm from the sample surface have enough kinetic energy to escape and be detected.

An XPS spectrum is a graph of the number of emitted electrons against the detected energies. Peaks are present in correspondence to the binding energies of the core electrons of the atoms contained in the material. As the area A_i of the peaks is proportional to the amount of atoms ejecting the electrons, the percentage atomic composition M_i of the sample can be calculated as follows:

$$M_i = \frac{A_i}{C_i} \times \frac{1}{\sum \frac{A_i}{C_i}} \tag{6}$$

where C_i is the photoionisation cross section for the atomic core electron i.

If the sample is not conductive, the ejected electrons can accumulate above the sample surface, and they can shift the position of the peaks eventually, making the peak determination impossible.

Measurements on flat samples.   The measurements were performed at the XPS laboratory, Institute of Physics of the University of Basel, in collaboration with Prof. P. Oelhafen.

The samples, with dimensions of approximately 10 mm × 20 mm, were fixed with metal screws on a metal holder and inserted into the spectrometer where an air pressure of about 10^{-9} mbar is reached in about 10 h. In spite of the high vacuum attained during the measurement, no damage was visible on the samples. The samples were then bombarded with X-ray and the spectra recorded in a few minutes.

Due to high surface-charging effects, it was possible to record spectra only from the silver mirroring areas close to the metal screws. In these areas, the surface charging was minimised because the electrons could diffuse to the metal holder and dissipate.

2.3.2. Size and shape distributions of the silver mirroring particles

The size and shape distributions of the silver mirroring particles, belonging to the edge mirroring present on a historical non-processed glass plate, was analysed, making use of transmission electron microscopy (TEM). The TEM micrographs were later digitised, and the digital images were analysed with image analysis software. Software from the Mathematica® libraries was used in the statistical analysis of the data.

2.3.2.1. Transmission electron microscopy (TEM)

Principle of the technique.   A TEM basically consists of a column of vacuum where electrons, emitted by a heated tungsten filament, are accelerated to a high voltage (200 keV) and then focussed by electron lenses (condenser lenses) on the specimen. The specimen is a thin section of material (thickness on the order of 70 nm) partially absorbing the electrons. The transmitted beam is then enlarged and focussed by electron lenses (imaging system), so that an enlarged image is formed on the viewing screen, and a plate covered with phosphors fluoresces when hit by the electrons. If a permanent image of the sample is required, the screen is removed and the electrons can directly hit a photographic film, which is subsequently developed and printed. For a more in-depth description of TEM, see Hayat (2000).

The electron absorption by the specimen is dependent on the sample thickness and composition (dense areas absorb more electrons). This determines an amplitude contrast and, therefore, an image.

As the electrons have to pass through the sample, the sample has to be 100 nm thick (maximum). Thin slices can be obtained by cutting the sample with an ultramicrotome. Normally, a soft sample cannot be cut straightaway but it has to be embedded in a harder material first. Sample embedding is a crucial factor in TEM.

Measurements.   Small portions (dimension 1 mm × 2 mm) of mirrored emulsion of a historical non-processed glass negative were removed by immersing the plate in a solution of water and alcohol, and using a knife blade and tweezers under the loupe. The samples were then embedded in a resin following a standard embedding method reported in Kejser (1995). The resin was a standard mixture of epoxy embedding medium, hardener (DDSA and MNA), and accelerator (BDMA), all of which were produced by Fluka. They were laid flat on a resin drop and, after removing the excess water on the emulsion, were covered with a second drop of resin and inserted into an oven at 60° for about 8 h. Later, slices of 70 nm thickness were cut in the cross-sectional direction with an ultramicrotome. The slices were transferred onto a grid and were then placed under the microscope. Images with magnification on the order of 10 000 times were taken. As the negative was a non-processed plate, long electron exposures had to be avoided in order to avoid the development of the silver bromide grains physically. The TEM micrographs were digitised, and the digital images were analysed with the software NIH Image®. The software Mathematica® was used in the statistical analysis of the data.

After embedding, slices of thickness 70 nm were cut in the cross-sectional direction with an ultramicrotome. The slices were transferred onto a grid, and they were placed under the microscope available at the Interdivisional Electron Microscopy Laboratory of the University of Basel (LEO EM 912), where images with magnification on the order of 10 000 were taken. As the negative was a non-processed plate, long electron exposures had to be avoided in order to avoid the development of silver bromide grains physically.

2.4. Results

2.4.1. Results on the chemical composition of silver mirroring

The results of the analysis of the chemical composition of silver mirroring are presented in Table 2. The XRD analysis on powder samples has determined that the silver mirroring scratched off from plates 1 and 2 is composed of 100% silver sulphide (Ag_2S). No other crystalline compounds were detected. In the case of detection of a single compound, the error in the mass composition of the sample attained by the instrument used in this measurement is on the order of 5–10%.

The XPS analysis revealed in sample 2c (Fig. 7) the presence of the following elements: carbon C (40%), oxygen O (11%), silver Ag (29%), sulphur S (14%), iodine I (4%), and mercury Hg (2%). The percentages refer to the relative atomic composition and are calculated from the spectra using formula (6). In sample 3b, carbon C (49%), oxygen O (10.5%),

Table 2. Results of the analysis of the chemical composition of silver mirroring

Sample	XRD powder	XPS
1a	Ag_2S	
2a	Ag_2S	
2c		Ag_2S
		O, C, I, Hg
3b		Ag_2S
		O, C, Br, I, Hg

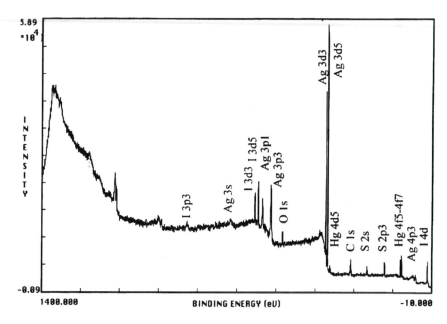

Fig. 7. XPS spectrum of sample 2c.

silver Ag (25%), sulphur S (11%), bromine Br (3%), iodine I (1%), and mercury Hg (0.5%) were detected.

In both samples, the main component is carbon, arising from the collagen contained in the gelatin emulsion. It is followed by silver and sulphur. The ratio between the amounts of silver and sulphur in the mirrored areas ranges between 2.07 and 2.27, almost stoichiometric for silver sulphide Ag_2S. Moreover the S(2p) peak has, in both cases, a negative shift (ranging from –4.5 to –6.3 eV), suggesting that sulphur is in the S^{2-} state (Hammond *et al.*, 1975). This indicates that silver sulphide is present in the mirroring areas. The amount of oxygen detected probably arises from the collagen present in the emulsion; indeed, the absence of the O(1s) peak at 529 eV excludes the presence of silver oxide (AgO) or silver dioxide (Ag_2O). The presence of bromine in sample 3b is consistent with the fact that plate 3 is a non-processed negative.

The XRD analysis on flat samples uniquely determined the presence of silver (Ag) crystal structures in sample 2b. In sample 3a, both silver and silver bromide (AgBr) structures were identified, consistent with the fact that plate 3 is a non-processed negative.

2.4.2. Results on the size and shape distribution of the silver mirroring particles

The samples were difficult to cut because of the softness of the specimen core. Although the slices often broke apart after cutting, it was possible to obtain some sections suitable for the analysis of particle distribution (Fig. 8).

The micrograph shown in Fig. 8 was digitised and used to analyse the variation of area, density, and sphericity of the small silver mirroring particles residing underneath the main layer. With the help of Adobe® Photoshop®, the upper silver mirroring particles were

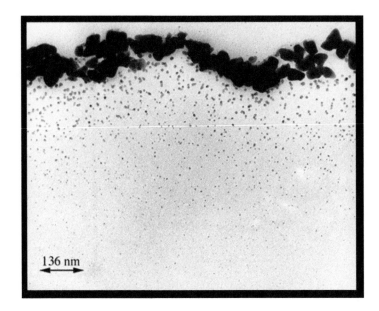

Fig. 8. TEM micrograph of the cross section of a mirrored area on a non-processed glass negative.

removed, and the image was divided into five strips. With the help of the NIH Image® software, each strip was converted into a binary (black–white) image (Fig. 9).

In each strip, the particles were counted and the particle mean area a (cm^2) and the length of the axes of the ellipse best fitting each particle were calculated. Figure 10 shows the resulting depth profile of the silver mirroring spatial particle density, while Fig. 11 shows the resulting depth profile of spherical and ellipsoidal particles. The percentage of spherical particles (star symbols) is almost constant in the emulsion, at about 60%. The particles with axes ratio bigger than two (triangle symbols) are always less than 20%, and show a slight increase in the uppermost 200 nm.

2.5. Discussion

The experiments have shown that the material contributing to silver mirroring is silver sulphide, and that the majority of the particles beneath the main silver mirroring layer have a spherical shape up to the area closest to the emulsion surface. This allows us to conclude that silver mirroring is due to the reaction between silver ions, a product of the oxidation of the image grains, with an environmental sulphur-containing compound, possibly hydrogen sulphide, and that this reaction takes place at the top surface of the emulsion. On the contrary, yellow discoloration takes place if the reactants are present within the emulsion. This happens either if they are the result of processing steps, so they are somehow formed along with the photograph, or if they can penetrate into the emulsion due to their high solubility and low reaction rates.

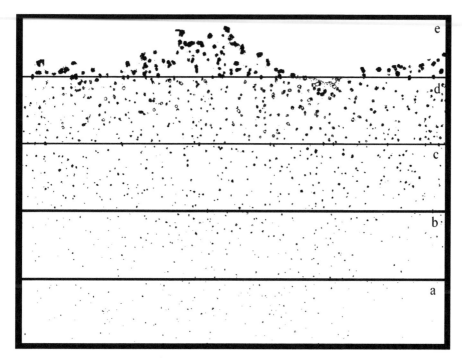

Fig. 9. Result of the division into five strips and of the binary conversion of Fig. 8.

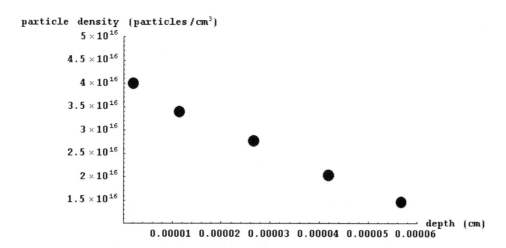

Fig. 10. Depth profile of silver mirroring particle density.

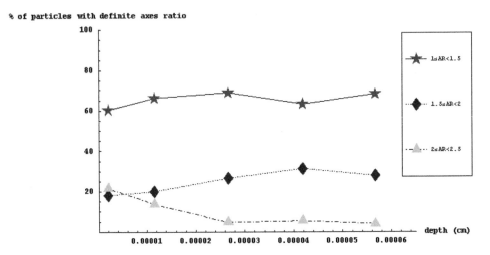

Fig. 11. Depth profile of the percentage of spherical (with ratio between the lengths of the ellipse axes between 1 and 1.5) and ellipsoidal particles.

The oxidation–migration–re-aggregation model for the formation of silver mirroring can be modified in a model consisting of the following steps: oxidation, diffusion of silver ions, reaction with external sulphur compounds, and growth of silver sulphide particles.

2.5.1. Oxidation

As in the classic oxidation–migration–re-aggregation model, the first step in the formation of silver mirroring is the oxidation of the image silver grains. This step has already been described in detail by other authors (Henn and Wiest, 1963; Brandt, 1987; Hendriks et al., 1991b). Here, I add only one comment relative to the case in which the oxidant compound is hydrogen peroxide.

For hydrogen peroxide partial pressures typically found in museums (on the order of ppb), the amount of hydrogen peroxide dissolved in an emulsion is on the order of 10^{-6} mol cm^{-3}, which is calculated assuming that the hydrogen peroxide is dissolved in the water contained in the emulsion and applying the Henry's law (7):

$$c_0 = p \times H^* \times \mathrm{mc} \times W_{H_2O_2} \tag{7}$$

where p is the partial pressure of hydrogen peroxide, H^* is the Henry's coefficient ($H^* = 1.8 \times 10^2$ mol cm^{-3} atm^{-1}), mc (–) is the moisture content of the gelatin, and $W_{H_2O_2}$ (g mol^{-1}) is the molecular weight of hydrogen peroxide ($W_{H_2O_2} = 34$ g mol^{-1}). The amount of hydrogen peroxide dissolved in the emulsion is two orders of magnitude lower than the amount of silver found in an emulsion on average, on the order of 10^{-2} g cm^{-3}, which corresponds to 2×10^{-4} mol cm^{-3}. This means that the oxidation step is always determined by the amount of hydrogen peroxide.

2.5.2. Diffusion of silver ions

The second step in the formation of silver mirroring is the diffusion of silver ions in the gelatin, driven by the difference in silver ion concentration between the areas closest to the image silver grains and the emulsion bulk. The diffusion of silver ions in water-soaked emulsions is very fast. Curtis and Leaist (1998) report that in wet gelatin gel, at room temperature, the diffusion constant of silver ions $D(Ag^+)$ is 1.6×10^{-5} cm^2 s^{-1}. However, silver mirroring is formed in normal museum conditions where the moisture content of the emulsion is only on the order of 20% maximum. From the work of Tanaka *et al.* (1973) on the conductivity K (ohm^{-1} cm^{-1}) and the silver ions transport number $\tau(-)$ in silver nitrate ($AgNO_3$) films kept at 79% RH and 25°C, the diffusion constant of silver ions can be calculated (Moore, 1972, Chapter 10, Section 13):

$$D(Ag^+) = \frac{RT}{F^2} \times \frac{K_{tot} \times \tau}{m(Ag^+)} \tag{8}$$

where R is the gas constant (R = 8.3143 J K^{-1} mol^{-1}), T (K) is the absolute temperature, F is the Faraday constant (F = 96 487 C mol^{-1}), K_{tot} (Ω^{-1} cm^{-1}) is the total film conductivity, and $m(Ag^+)$ (mol cm^{-3}) is the molar concentration of the silver ions in the film.

By inserting Tanaka's data for τ, K_{tot}, and $m(Ag^+)$ in equation (8), the diffusion constant of silver ions in gelatin films kept at 79% RH and 25°C becomes of the order of 5×10^{-11} cm^2 s^{-1}.

Assuming a diffusion law of the kind

$$\Delta L = \sqrt{D \times \Delta t} \tag{9}$$

where ΔL (cm) is the distance travelled in a time Δt (s) by species with diffusion constant D (cm^2 s^{-1}), silver ions in emulsions at 79% RH and 25°C travel 50 μm (typical emulsion thickness in glass plates) in about 6 days. This time is short in comparison with the observed timescale of silver mirroring formation under normal archive conditions, which is on the order of months or few years maximum. This allows us to conclude that, in typical archive conditions, silver ions diffuse rather fast in the emulsion, and they can be responsible for the formation of silver mirroring.

No data have been found in the literature about the diffusion constant of colloidal particles in gelatin. Nevertheless, it is possible to estimate this by observing that the image in silver gelatin printed out papers (POPs) dating back to the beginning of the twentieth century is made of silver colloidal particles of radius on the same order of magnitude as the particles found beneath the silver mirroring layer, *i.e.* on the order of 5 nm (Lavedrine, 1991). As these images are still very sharp, particles of this size do not cover a distance sufficient to have a blurred image (assumed to be 0.1 mm (= 10^{-2} cm)) in 100 years ($\approx 3 \times 10^9$ s). By applying equation (9) with $\Delta L = 10^{-2}$ cm and $\Delta t > 3 \times 10^9$ s, the diffusion constant of these particles in gelatin becomes smaller than 3×10^{-14} cm^2 s^{-1}. Therefore, these particles would need more than 25 years to cover a distance of 50 μm, the typical emulsion thickness, a time much longer than the typical time of formation of silver mirroring.

This allows us to conclude that the formation of silver mirroring is due to the diffusion of silver ions and not of colloidal particles in the emulsion.

2.5.3. Reaction with external sulphur compounds

Once silver ions are produced and diffuse homogeneously in the emulsion, they will react with external sulphur-based compounds to produce silver sulphide seeds. Among environmental gases, the most probable gas is hydrogen sulphide (H_2S) (typical sources of hydrogen sulphide in archives or museums are natural fibres, wool, humans, and rubbers), a second possibility is carbonyl sulphide (OCS).

To get an insight into the detailed reaction mechanism between hydrogen sulphide and silver ions, it is useful to look at the studies on the formation of silver sulphide on silver plates, a phenomenon usually called silver tarnishing.

These studies have pointed out that if moisture (Lilienfeld and White, 1930; Pope et al., 1968; Bennett et al., 1969; Reagor and Sinclair, 1981; Franey et al., 1985; Graedel et al., 1985; Volpe and Peterson, 1989) or oxidant gases (Volpe and Peterson, 1989) are present, the tarnishing rate increases. In this case, the rate-limiting step is not the reaction but the time needed for the gas to reach the plate, the so-called mass transport rate in air (Reagor and Sinclair, 1981; Volpe and Peterson, 1989). Moreover, it has been observed that silver sulphide growth occurs on silver sulphide clumps created at the initial exposure (Bennett et al., 1969; Franey et al., 1985). Once the clumps coalesce in a continuous film, the tarnishing rate is limited by the diffusion of the silver ions through this film.

Although the increase in the reaction rate in the presence of moisture is widely recognised, no agreement is found on the detailed role of moisture in the reaction. Three possible pathways involving water are given:

1. Oxidation of H_2S by oxygen in water to give sulphur, followed by direct reaction of sulphur with silver (Lilienfeld and White, 1930; Volpe and Peterson, 1989).
2. Dissolution of H_2S to HS^-, followed by direct reaction of HS^- with silver (Graedel et al., 1985) and
3. Absorption of H_2S in water and direct dissociative co-ordination of H_2S with silver resulting in the formation of an unspecified intermediate. This intermediate reacts with a second silver atom to give silver sulphide (Graedel et al., 1985).

The previous observations suggest that, in silver mirroring, the reaction rate between hydrogen sulphide and silver ions is fast in comparison with the diffusion rates. Indeed, the reaction can be assumed to take place in a water environment because photographic emulsions contain about 20% of water by weight. In addition, oxidised emulsions contain hydrogen peroxide, which is able to oxidise not only the silver grains but also hydrogen sulphide.

It is possible to envisage the following pathway for the dissolution of hydrogen sulphide in water followed by the reaction with silver ions:

(a) $H_2S \rightarrow H^+ + HS^-$
(b) $HS^- \rightarrow H^+ + S^{2-}$
(c) $2\,Ag^+ + S^{2-} \rightarrow Ag_2S$

Although it has not been possible to calculate the reaction rate, it is important to notice that the constant of dissociation of (a) is $k_a = ([H^+] \times [HS^-])/[H_2S] = 9.8 \times 10^{-8}$ and of (b) is $k_b = ([H^+] \times [S^{2-}])/[HS^-] = 1.1 \times 10^{-12}$. Therefore, hydrogen sulphide is completely dissociated into S^{2-} only for pH greater than 12. Nevertheless, as the solubility constant of silver sulphide is very low ($k_{sp} = [Ag^+]^2 \times [S^{2-}] = 6.89 \times 10^{-50}$), precipitation of silver sulphide will occur as soon as S^{2-} ions are in solution.

Silver sulphide seeds are produced at the emulsion upper surface for three reasons. First of all, the upper surface is the region where the reactants, silver ions present in the emulsion and hydrogen sulphide present in the atmosphere, first meet. Calculations of the concentration profiles for a similar problem (S^{2-} ions that penetrate into a gel containing a second reactant Pb^{++} and form a precipitate at the interface) were performed by Hermans (1947). Second, hydrogen sulphide is extremely less soluble in water than hydrogen peroxide. The Henry coefficient H^* for hydrogen sulphide is 9.8×10^{-2} mol l^{-1} atm^{-1} (Fogg and Gennard, 1991), seven orders of magnitude smaller than the Henry coefficient for hydrogen peroxide. This implies that for partial pressures of hydrogen sulphide typically found in museums, on the order of ppb, the amount of hydrogen sulphide dissolved in the gelatin is on the order of 10^{-13} mol cm^{-3}. In this case, the reaction between silver ions and hydrogen sulphide is controlled by the amount of hydrogen sulphide. This, added to the fact that the rate of reaction is probably faster than the rate of hydrogen sulphide diffusion into the gelatin, results in a lesser penetration of hydrogen sulphide into the emulsion.

2.5.4. Growth of silver sulphide particles

The final step in the formation of silver mirroring is the growth of silver sulphide particles. The silver sulphide seeds grow because of the reaction between silver ions and hydrogen sulphide molecules. The reaction does not have a preferential orientation; therefore, the final shape of the particles is spherical. Further exposure to hydrogen sulphide will provoke the growth of the seeds without increasing their number. This is supported by two different types of studies found in the literature. The first type of studies is concerned with the tarnishing of silver plates. Bennet et al. (1969) have shown that silver sulphide clumps on silver plates nucleated on initial exposure to hydrogen sulphide and that further reaction occurred on the initially formed clumps. Graedel et al. (1985) also report the same reaction dynamics.

The second type of studies is related to the photographic processes called diffusion transfer processes (typically used in Polaroid photographs). In these processes, silver sulphide particles are used to catalyse the reaction between the developer and the silver ions (James, 1939; Eggert, 1947; Shuman and James, 1971; Levenson and Twist, 1973) either by adsorbing the developer onto the colloidal particles, or by stabilising a single silver atom using the electrical conductivity of the colloidal particle. The stabilisation of silver atoms by silver aggregates has also been the object of more recent studies.[2] Although diffusion transfer processes differ from silver mirroring because the reaction takes place between silver ions and hydrogen sulphide instead of between silver ions and developer, colloidal silver sulphide particles could play the same catalytic function.

The difference in size between the particles at the emulsion–air interface and the particles underneath can be explained because the particles at the interface grow relatively fast as they are directly exposed to the environmental hydrogen sulphide. The more they grow, the more they fill the emulsion surface, hindering the penetration of the gas into the emulsion. When the surface is completely covered, the amount of hydrogen sulphide entering the emulsion is zero, and the growth of the particles underneath the surface is blocked.

---

[2] It has been shown (for a review see Henglin, 1993) that the electrochemical potential for the reaction, $Ag^+ + e^- \rightarrow Ag^0$, is very low for single ions (−1.8 V) and it increases with the size of the silver aggregates till the value assumed on the solid metal (+ 0.799 V). For small silver aggregates, quantum effects have been taken into account (Belloni et al., 1991), while for aggregates larger than 1 nm a simple surface energy effect explains this behaviour (Plieth, 1982).

2.6. Conclusions

New spectroscopic experiments on the chemical composition of silver mirroring on silver gelatin glass plates have shown that silver mirroring is composed of silver sulphide (Ag_2S). Transmission electron micrographs of cross sections of the mirrored region on a non-processed glass plate have confirmed the results of Nielsen and Lavedrine (1993): silver mirroring is formed on a surface layer of closely packed particles of dimensions on the order of 100 nm, underneath where a large number of smaller particles of dimensions on the order of 10 nm are found. The analysis of the size and shape distribution of these particles has revealed that the majority of the particles, although their size increases with their proximity of the emulsion surface, are spherical.

Based on these results and on theoretical calculations on the diffusion of ions and particles in the gelatin, some modifications to the oxidation–migration–re-aggregation model for the local formation of silver mirroring are proposed. The modifications are mainly concerned with the role played by silver ions, with the chemical reactions leading to silver sulphide particles at the emulsion surface and with the mechanism of growth of the silver mirroring particles.

They allow the prediction of the conditions under which yellow discoloration and not silver mirroring takes place. Indeed, colloidal particles are formed in the emulsion, and they result in yellow discoloration if the reactants are present in the emulsion bulk. This happens either if they are the result of processing steps or if they can penetrate into the emulsion due to their high solubility and low reaction rates. Such particles are immobile in the gelatin. On the other hand, this work has shown that silver mirroring is the result of the reaction of silver ions, mobile in the gelatin, with external sulphur-containing compounds.

In addition, this work has shown that the total rate of silver mirroring formation does not depend on the mass transport rate of the gases and silver ions in the emulsion and that, as long as the silver mirroring particles do not fill the emulsion surface, the reactions are under the control of the amount of the external oxidant and sulphur-containing gases.

The model presented here deals with the interaction of gases with the photographs, but it would not substantially change in case oxidising and sulphur-containing compounds were present in the material directly in contact with the photograph surface.

3. IDENTIFICATION OF PHOTOGRAPHIC DYES IN COLOUR MOTION PICTURE FILMS

Shortly after the invention of photography (in the 1830–1840s), motion picture films were invented (in the 1890s) and soon became popular. Nowadays, film archives worldwide store impressive amounts of motion picture films that tell the story of the twentieth century life and customs. So far, the most urgent problem for film preservation is the deterioration of the nitrate or acetate support. In the last twenty years, the understanding of the degradation mechanism of cellulose nitrate and cellulose acetate has improved considerably (Adelstein et al., 1992a,b, 1995a–c) and has led to the development of international standards for the preservation of photographic materials on nitrate and acetate base. Another widespread problem for film preservation is the fading of the dyes in colour motion picture films.

As a result of the fading of the dyes, the film acquires an overall hue, for example, magenta. The original colours cannot be restored chemically, and the only solution available at the moment is to digitise the film and reconstruct the colour on a digital basis.

Dye degradation can occur as a result of exposure to light or, in the dark, as a result of heat, humidity, or environmental pollutants. As films stored in archives spend the majority of their life in the dark, this work was focussed on the understanding of the mechanism of dark fading. In particular, the long-term aim is to understand the relation between dye fading and the degradation of the base.

The susceptibility of dyes to degradation heavily depends on the precise chemical nature of the dyes and the structure of the layers in the film. In order to have some idea about the mechanism of dark fading of the dyes, it is necessary first of all to have a clear picture of the historical development of the dyes used in colour motion picture films. The major difficulty in tackling this problem is that the photographic industry will not reveal the nature of the dyes incorporated in the films, even many years after the discontinuance of production.

It is nevertheless possible to make some relevant progress in this direction. It is necessary, at first, to have a clear view of the different films produced by the photographic industries. Every time a relevant technological development happened, a new film stock was released, usually with a different name. The history of the film stocks produced by Kodak is presented in Section 3.1. Second, it is possible, from the literature, to trace a generic history of the dyes used by the photographic industries and the main mechanisms of deterioration of such dyes. This is presented in Sections 3.2 and 3.3. Finally, it is possible to select some spectroscopic techniques that are able to characterise the dyes, and progresses in this direction are presented in Section 3.4.

3.1. Evolution of Kodak colour film stocks

The starting point for tracing the historical development of dyes in colour motion picture films is to look at the evolution of the colour film stocks. In this work, the attention is focussed on Kodak films because they represent the majority of the films owned by the National Film and Sound Archive in Australia.

By collecting data from the chronology of films available on Kodak website (http://www.kodak.com/US/en/motion/about/chrono1.shtml) and from the Kodak Film Code History (a spreadsheet owned by Kodak Australia containing more than 250 entries), we produced a timeline for Kodak colour motion picture films. This is a matrix of date and film stock representing the colour motion picture films produced by Kodak in the time range 1935–2004. The timeline and notes for each film stock give an idea about the extent of technological change introduced in colour films in the last 50 years. In particular, between 1965 and 1985, the period in which major advances in dye stability were obtained, Kodak produced about 12 different negative films, 19 different print films, and 16 different intermediate films. The complete timeline for negative, positive, and intermediate films in the period 1935–2004 was published elsewhere (Di Pietro, 2005). At the National Screen and Sound Archive in Australia, more work is at present being done to explore the possibility of identifying each of these film stocks through the notch codes and to assess the colour stability of the film by colorimetric measurements on the section of films where only the dyes in the masking layer are present.

3.2. Types of dyes used in colour motion picture films

Only those photographic processes where coloured dyes are formed by the reaction of the oxidised colour-developing agent (*p*-phenylenediamine derivatives) with colour couplers present in the emulsion, the so-called chromogenic processes, are reviewed here. These are the most common photographic processes used in colour motion picture films. The colour couplers are incorporated into separate layers in the emulsion to form magenta, cyan, or yellow layers. Magenta and yellow dyes are of the azomethine family, while the cyan dyes are of the indoaniline family. Generally, a print film will have the magenta-forming layer on the top of the emulsion, followed by the cyan and the yellow. In contrast, a negative film will have the yellow on the top followed by the magenta- and cyan-forming layer. Each layer has a thickness between 5 and 10 μm.

Over time, the chemical nature of colour couplers has undergone many changes. This has been driven by the necessity to find couplers giving better hues, with more stability, with improvements in grain size and sharpness, and at a lesser cost. Different couplers are used in different film stocks, depending on the use of the film (negative, print, reversal film, and intermediate).

The colour couplers are generally divided into 2 or 4 equivalent couplers, depending on how many silver halide molecules have to be consumed to form one dye molecule. Different groups can be attached to the main coupler body to prevent the diffusion of the couplers. These are either hydrophobic groups (the resulting dyes will have a micelle structure), polymeric chains chemically bonding to the couplers, or constituents capable of forming insoluble salts with heavy metals or ballasting non-hydrophilic groups (in this case, they are dissolved in a solvent and dispersed in the emulsion mechanically). The precise nature of the final dye will depend on the nature of the coupler, the nature of the ballasting groups, and the nature of the developing agent (see Fleckenstein, 1977). To understand the complexity of this problem and the number of different colour couplers used in the last 50 years, see Bergthaller (2002a–c). This review, however, refers mostly to couplers developed by Agfa.

The most important classes of *yellow-forming couplers* are α-pivaloyacetanilide and α-benzoylacetanilide types (Fig. 12). The chemical reaction that gives rise to the yellow dye is described in Theys and Sosnovsky (1997). Historically, benzoylacetanilides are the prototypes of yellow couplers for their high tinctorial strength, but they are not very stable. Pivaloyacetanilides couplers are more stable and were patented in 1966 from Kodak. In 1993, Fuji patented a new type of yellow couplers that are more stable and have higher tinctorial strength, the cycloalkanoylacetanilides (Bergthaller, 2002a).

The most important classes of *magenta-forming couplers* are 5-pyrazolinone, indazolone, pyrazolobenzimidazole, and pyrazolotriazol (Fig. 13). The chemical reaction that gives rise to the magenta dye is described in Theys and Sosnovsky (1997).

Until 1980, pyrazolones were the magenta couplers of choice; in particular, the 3-acylaminopyrazolones, which were discovered before 1950 (Fig. 14). However, they are sensitive to aerial oxygen. Prolonged storage may result in air oxidation and, in contact

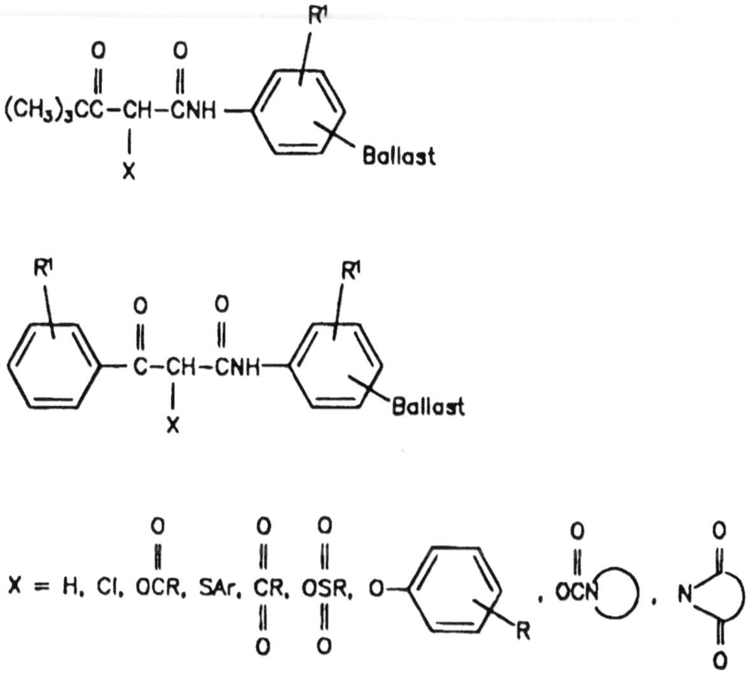

Fig. 12. Yellow couplers. α-Pivaloyacetanilide type (top) and benzoylacetanilide type (bottom). R and R^1 are various organic moieties, and ballast can be various long-chain aliphatic groups (from Theys and Sosnoysky, 1997). Reprinted with permission from: Theys, R.D., Sosnovsky, G., 1997. Chemistry and color photography. *Chem. Rev.* **97**, 83–132. © 1997 American Chemical Society.

with aldehyde vapours (emitted from fibre board), they lose their coupling activity, which results in uneven magenta staining. The pyrazolotriazole couplers have improved thermal stability and lower side absorption as compared to the pyrazolinone couplers.

All *cyan couplers* are substituted phenol- or naphthol-type couplers (Fig. 15). The chemical reaction that gives rise to the cyan dye is described in Theys and Sosnovsky (1997).

3.3. Mechanisms of dye degradation in the dark

In this work, attention is focussed on the mechanisms of dye degradation in the dark. This is the most probable degradation mechanism because the motion picture films stored in the archive are seldom exposed to light. Information for this review was gathered from Tuite (1979), and the review articles of Bergthaller (2002c) and Theys and Sosnovsky (1997). The 1979 Tuite review article is, incredibly, the last published review article fully devoted to photographic dye degradation.

There are a number of reactions occurring in the dark that lead to dye fading. These involve either the dyes themselves or the residual couplers. Indeed, in the chromogenic

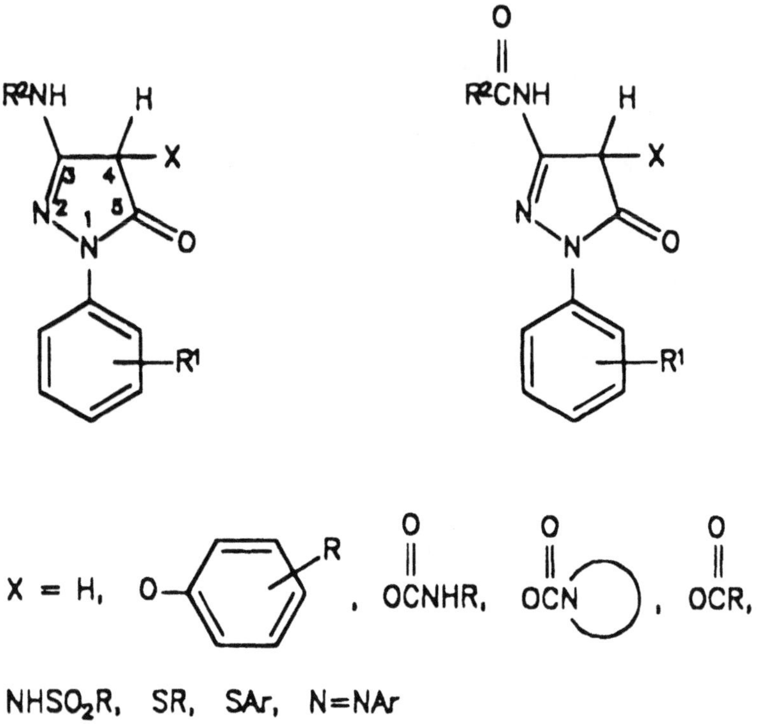

Fig. 13. Two couplers of the pyrazolinone class: 3-arylamino substituted (left) and 3-acylamino substituted (right). R, R^1, and R^2 are various organic moieties (from Theys and Sosnoysky, 1997). Reprinted with permission from: Theys, R.D., Sosnovsky, G., 1997. Chemistry and color photography. *Chem. Rev.* **97**, 83–132. © 1997 American Chemical Society.

X = H, Cl, SR, OR
R, R^1 = alkyl, aryl

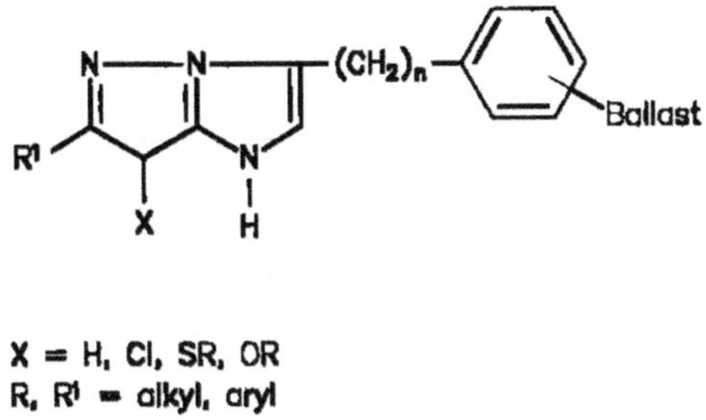

Fig. 14. Magenta-forming pyrazolotriazole couplers (from Theys and Sosnoysky, 1997). Reprinted with permission from: Theys, R.D., Sosnovsky, G., 1997. Chemistry and color photography. *Chem. Rev.* **97**, 83–132. © 1997 American Chemical Society.

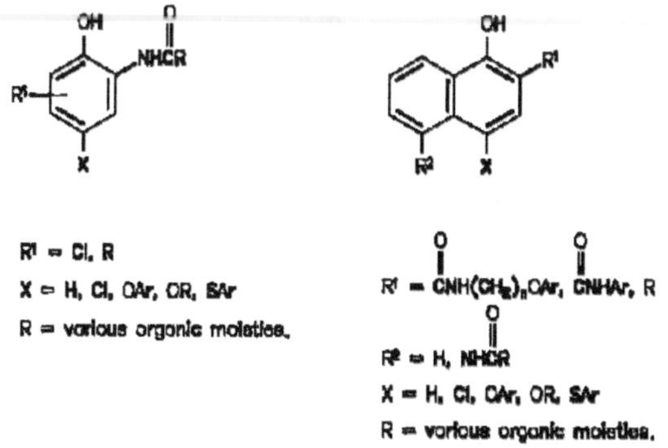

Fig. 15. Cyan-forming phenol (right) and naphthol (left)-type couplers (from Theys and Sosnoysky, 1997). Reprinted with permission from: Theys, R.D., Sosnovsky, G., 1997. Chemistry and color photography. *Chem. Rev.* **97**, 83–132. © 1997 American Chemical Society.

photographic processes, the colour couplers are ballasted in the emulsion and are not washed away after processing the film. This means that the unused couplers will stay in the emulsion and can undergo reactions leading to colour products. This is the reason why chromogenic processes (processes in which the colour couplers are not incorporated into the emulsion but are precipitated in the three layers during the processing steps, for example, Kodakchrome) have traditionally a high dark stability.

An important reaction related to magenta dye instability is the reaction involving the residual magenta coupler shown in Fig. 16 and resulting in the yellowing of the non-image areas. This reaction leads to a yellow-coloured compound, either in the dark or in the presence of light. The product is methynylbis coupler in the dark and the azo-dye in the light. This problem was solved with the substitution of two of the ortho positions of the 1-phenyl ring (Tuite, 1979).

Another important reaction occurs between the residual magenta coupler (3-acylamino pyrazolone couplers) and the magenta dye to form a variety of colourless products. Using aldehydes in the final processing bath solved this problem. Indeed, the cross-linked couplers do not react with the magenta dye. Later, magenta couplers (anilino pyrazolone) were devised, which showed much lower tendency to react with their magenta dyes and were used in paper prints (Tuite, 1979). Nevertheless, the 3-acylamino pyrazolone couplers were the most common couplers used in camera-use films such as colour negatives, colour reversal films, and colour films for motion pictures at least up to 1988 (Sakanoue and Furutachi, 1988) because of their image quality. Sakanoue and Furutachi (1988) found that the degree of dye decomposition relates to the pKa value of the coupler and that the use of low-pKa couplers could eliminate the necessity of formaldehyde from the processing solution. They proposed a fading mechanism, described in Fig. 17.

A third important class of reactions involving magenta dyes is the reaction with residual thiosulphate (from the fixing bath) (Miyagawa and Shirai, 1985; Kurosaki *et al.*, 1988). In these articles, the authors report the reductive fading pathway, the structures of the dye, and the colourless adduct.

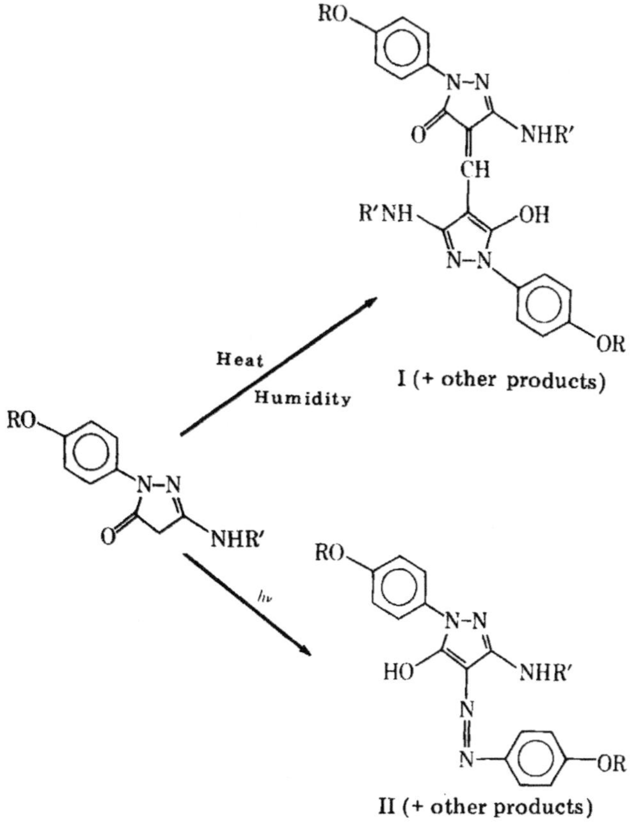

Fig. 16. Reactions involving residual magenta coupler (from Tuite, 1979). Reprinted with permission of IS & T: The Society for Imaging Science and Technology sole © owners of the Journal of Applied Photographic Engineering.

The result of the innovations on magenta couplers is that none of the modern (post-1979) Kodak colour photographic materials that use traditional processing are magenta dye limited for dark stability. They can be limiting for light stability. Adding an UV absorbing layer in the prints in front of the magenta layer circumvents this problem.

The *yellow dye* is the least stable dye in the dark because of hydrolytic reactions occurring both in acidic and alkaline environments. The acidic hydrolytic attack occurs at the azomethine linkage, while alkaline hydrolysis occurs at the keto linkage (Tuite, 1979). The final processing pH of a film is chosen to minimise these two reactions. The reaction shown in Fig. 18 is considered the most important dark fading reaction occurring when a film suffers from vinegar syndrome.

Indeed, yellow fading was monitored for years at Kodak as an indication of the onset of the degradation of the cellulose triacetate base (vinegar syndrome). Tuite (1979) reports that the dark fading characteristic of the yellow dye for 10% dye loss ranges from 3 years for Kodak Ektachrome 40 Movie Film (type A, EM-25 process) to 32 years for Kodak

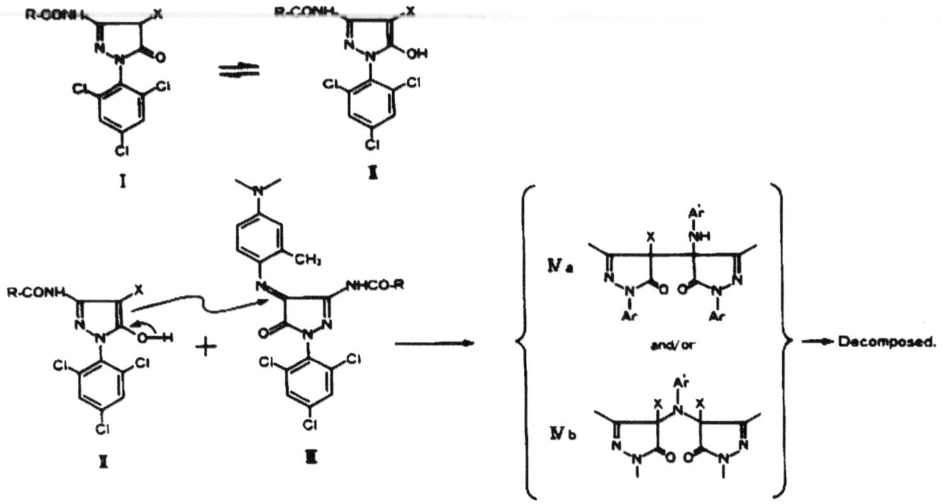

Fig. 17. Proposed mechanism for the reaction between magenta 3-acylamino pyrazolone coupler and magenta (from Sakanoue and Furutachi, 1988)

Ektachrome 50 Professional Film 6118 (Daylight, E-8 process). As far as light stability is concerned, the largest single improvement came from the simple change from a benzoyl group to a pivalyl group on the yellow coupler (Tuite, 1979).

The *cyan dye* is the dye most stable to light. The principal undesired dark reaction is the reduction of the residual cyan coupler to the colourless leuco form of the dye (Fig. 19). Other reducing agents that have not been removed completely can also cause this or similar reactions: *e.g.* retained thiosulphate from the fix bath, ferrous ion from a poorly regenerated bleach fix bath, and ballasted hydroquinones that are used as incorporated interlayer scavengers.

This problem was solved by a change in the colour-developing agent from CD-2 to CD-3 and by a change in the dye structure, namely the use of phenol-type couplers with amide group in the 2.5 position of the ring (Tuite, 1979; Theys and Sosnovsky, 1997).

The masking dyes that gives the orange colour to negatives is very stable in the dark (Bergthaller, 2002c).

Photochemical degradation involves absorption of light by a dye to generate both singlet- and triplet-excited states. Aerial oxidation of the excited molecules forms undesirable by-products. Stabilisation of the dye involves intercepting both the actinic radiation with a UV absorber and one or more of the reactive intermediate with a quencher such as nickel dibutyldithiocarbamate. Bergthaller (2002c) proposes a degradation trail common for both dark fading and photo fading. He assumes that under appropriate conditions the detachment of the developer fragment from a dye cloud could be started more readily from a sequence of single electron transfer and proton transfer to the azomethine bond. This idea is based on the fact that, in many cases, increased resistance of dyes to dark stability has resulted from efforts directed towards improved light stability.

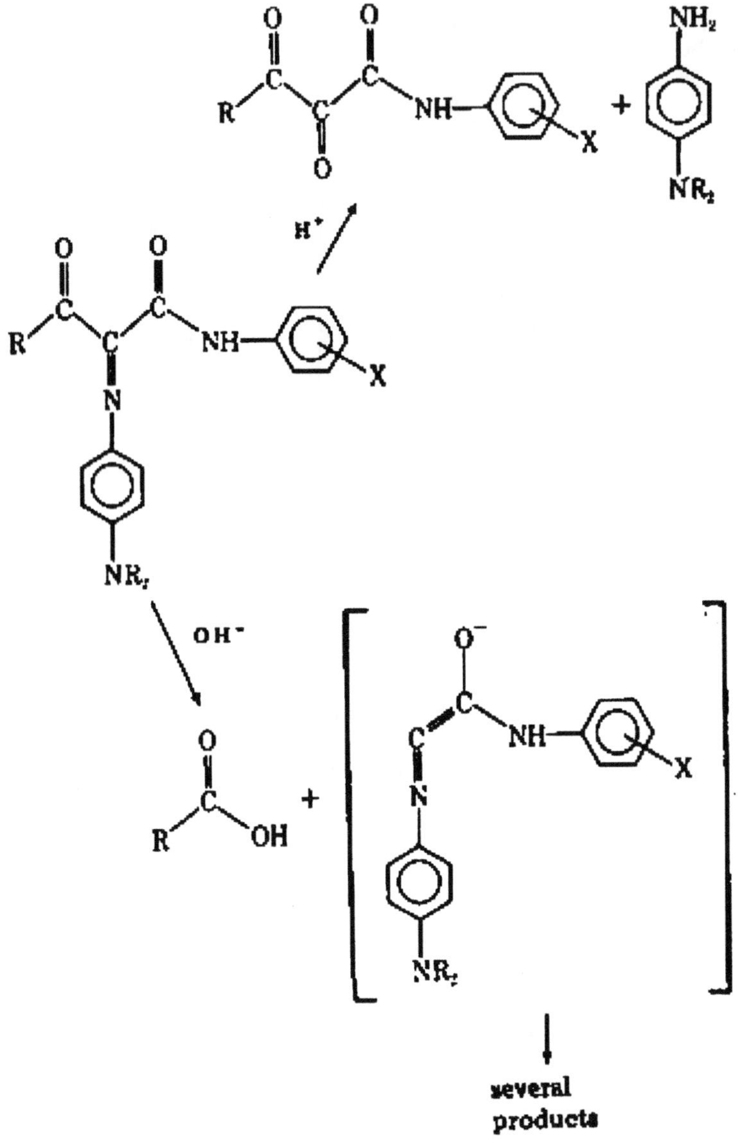

Fig. 18. Yellow dye hydrolysis under acid (top) and alkaline (bottom) conditions (from Tuite, 1979).

The atmospheric constituents, which most often cause colorants to fade, are oxides of sulphur and nitrogen, and ozone. Atmospheric oxygen can also be a significant contributor to the fading of certain types of yellow and magenta dyes. In contrast, it was found that oxygen could inhibit the fading of cyan dyes (Theys and Sosnovsky, 1997).

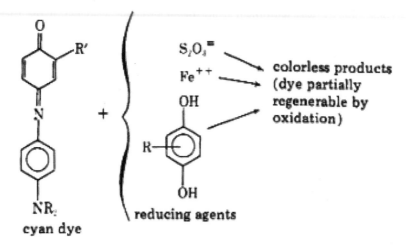

Fig. 19. Reaction between cyan dye and reducing agents (from Tuite, 1979). Reprinted with permission of IS & T: The Society for Imaging Science and Technology sole © owners of the Journal of Applied Photographic Engineering.

3.4. Experimental analysis of dyes

3.4.1. Choice of spectroscopic technique and sample preparation

A colour film has a layered structure, where each layer contains only one type of dye. Figure 2 shows the cross-sectional structure of typical print and negative stocks. Each layer has dimensions in the range of 5–10 μm. The "China girl" image in the leader of the movie film was chosen as the appropriate test film because it is closest to the surface and, therefore, more susceptible to environmental conditions, and it is on all the films, and it is not part of the film content.

To analyse the dyes present in each layer of the "China girl" image on the film leader, it is possible to follow three different strategies:

- Use an analytical technique that can select the layer non-destructively, *e.g.* using a confocal microscope associated with an FTIR spectrometer.
- Cut the film in the cross section and analyse the layers. This is achievable if the analytical technique has a detection area of diameter of 5 μm or less.
- Dissolve the dyes and separate them either with thin layer chromatography (TLC) or high-performance liquid chromatography (HPLC). The dyes can then be analysed with Fourier transform infrared spectrometry (FTIR), mass spectrometry (MS), or with other techniques.

The *first* strategy was attempted with the Raman Confocal Microscope at the University of Canberra. This was unsuccessful because the surface of the film gave a very high fluorescence signal that masked the signal from the underlying layers.

The *second* strategy was performed using the microtome available at the Electron Microscopy Unit of the Australian National University. This does not attempt to investigate the cross section of the film directly. A small piece of film (approximate dimensions of 2 mm × 2 mm) was cut with a scalpel and mounted flat on a resin capsule using superglue. Care was taken to stick the film base to the resin capsule and not the film emulsion side. The capsule was then fixed on the microtome, and thin sections were cut with a glass knife as precisely as possible parallel to the base. As it is not possible to have sections parallel to the base on the whole film surface, the cut will be slanted, and it will expose the three colour layers (Figs. 20 and 21). The slanted section can be used for analysis and imaging.

The thin sections can be either discarded or kept for further analysis. In the second case, it is important to collect specimens continuously. This can be achieved by floating them in isopropyl alcohol. This solvent is contained in a so-called boat built and attached on purpose to the glass knife. It is difficult to retrieve thin sections flat because they tend to curl up. It is essential that the boat be completely full and the sections (which tend to sink) be lifted using a small loop made by a hair (human or animal) attached to a bamboo stick.

TLC coupled with FTIR has been used at the Research School of Chemistry of the Australian National University. To separate the dyes with TLC, it is necessary to dissolve them, leaving a piece of film in a vial with ethanol for few days. The dissolution process is aided by little heating and stirring. The ethanol is then allowed to evaporate, and diethyl ethanol is added to the vial. Diethyl ethanol is used because of the higher solubility of the dyes in this solvent. It cannot be used initially because of its high evaporation rate. A few millilitres of this solution is spread following a line on a large silica TLC plate of dimension 200 mm × 200 mm (DC-Alufolien Kieselgel60 produced by Merck) using a glass capillary. The use of large TLC plates is necessary if sufficient compounds are to be collected for further analysis. The TLC plate is then immersed in a tray containing the mobile phase (100 ml of a solution of diethyl ether and petroleum spirit in diethyl ether, in proportion 40:60 for the prints and 90:10 for the negatives) and left there for the time necessary for the mobile phase to rise to the top of the plate (~90 min). This procedure produces a plate with separate lines of the different dyes. It is noticeable that fresh print film gave rise to 3 lines (magenta, cyan, and yellow), while aged print films gave rise to more lines (usually the three main colour lines followed by a second much fainted line of the same colour). This is due to the fact that ageing has caused breakdown of some of the dye molecules. The dyes can be scratched from the plate and dissolved in diethyl ether. After the silica is decanted, the remaining liquid is dried and mixed with KBr (potassium bromide) to produce pellets that can be analysed with FTIR. Unfortunately, the fainter lines, which probably contain the dye degradation products, did not yield sufficient material to produce a pellet and could not be analysed with FTIR. The dyes were analysed using Raman spectroscopy (RS) and FTIR.

3.4.2. Raman spectroscopy

The Renishaw 2000 Raman Microscope at the University of Canberra was used with the laser beam perpendicular to the film surface, in an attempt to use the confocal characteristics of the microscope and to detect signals from the three different layers.

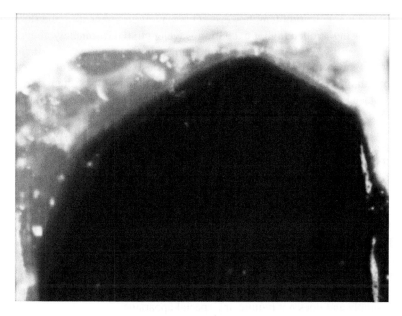

Fig. 20. Reflection light microscopy photograph (×64) of the slanted section of the Kodak print film 5384 (~1980). Notice the black-coloured top layer (resulting from the superposition of the magenta cyan and yellow layer) and the exposed cyan and yellow layers.

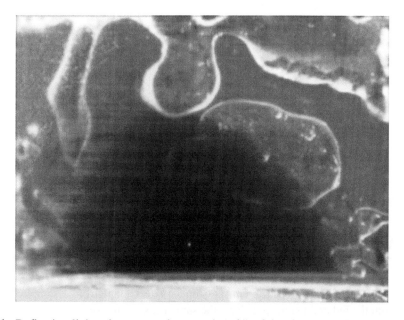

Fig. 21. Reflection light microscopy photograph (×64) of the slanted section of the Kodak print film 5384mod (or 5386) (~1995). Notice the exposed cyan and yellow layers.

Fluorescence remained a problem, however. One way of circumventing the fluorescence problem is by using a much higher excitation wavelength. Unfortunately, the tests made using the Bruker IFS 100-Fourier transform Raman spectrometer with excitation wavelength of 1064 nm at the University of Sydney were also not successful, because of the limited optical throughput of the instrument. In this geometry (the system does not have a microscope and requires much higher sample masses), the signal was dominated by the cellulose acetate base, and the dyes were not visible (Fig. 22). It is important to notice that Fourier transform Raman spectroscopy has been used successfully to identify photographic dyes present in effluent streams. In this case, the dyes are first concentrated using solid-phase extraction (SPE) and then analysed in the FT-Raman apparatus (Bristow *et al.*, 1998).

The Renishaw 785 nm Raman microscope was then used to analyse the cross section (Fig. 23). Cross section geometry has two advantages: it maximises the scattering volume, and it allows the spatial separation of the three coloured layers. Cross sections can be produced either with the methods explained earlier or by briefly immersing the emulsion in liquid nitrogen and then cracking and tearing it. This causes the emulsion to become delaminated from the base. The former procedure (microtoming) is far more reproducible and precise than the latter. Still, the delamination procedure has the advantage of being quick and simple and does not require any special apparatus.

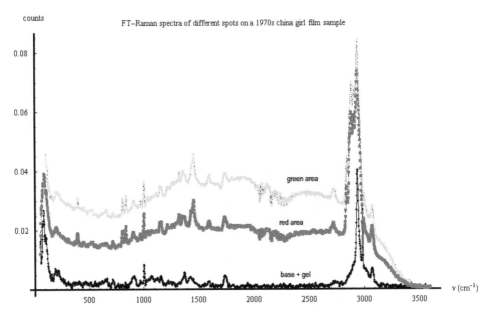

Fig. 22. FT Raman spectra of different spots on a 1970s China girl film sample. The spectra from the red and green spots have the same peaks as the spectra from the cellulose acetate base and gelatin spot.

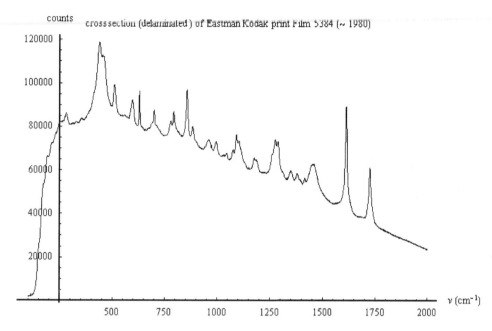

Fig. 23. Raman spectra of the cross section (delaminated) of Eastman Kodak print film 5384 (~1980).

A pronounced fluorescence shoulder still dominates the spectra, but characteristic Raman peaks are also visible. The fluorescence shoulder can be removed by background subtraction.

A more elegant way to get rid of the fluorescence signal is to use a technique dependent on the subtraction of "fixed pattern non-uniformity" and known as subtracted shifted Raman spectroscopy (SSRS) (Bell *et al.*, 1998). The basic idea comes from shifting the spectrum of a sample by a small amount of about 15 cm⁻¹, by changing the angle of the diffraction grating in the spectrometer. This overcomes the noise-limiting non-uniformity of adjacent pixels in a CCD detector. The two spectra are subtracted, and the Raman peaks appear as derivative waveforms. Curve fitting is then applied, and the original Raman spectrum is reconstituted without the fluorescence background.

The microtomy method explained earlier has not only the advantage of being repeatable and more precise, it also offers the possibility, through the slanted cut, to expose an area of a single layer that is large enough (about 10 μm) to be independently detected in the Raman microscope. Figure 24 shows the Raman spectra of each single layer in the Kodak print film 5384.

This was compared with spectra from a recent Kodak film stock (5384 or 5386). As Fig. 25 shows, the two spectra are different, a proof that RS can identify the differences in dyes. Unfortunately, in both cases, the spectra from the cyan and magenta layer do not show different characteristic peaks, and the spectra of the yellow layer are dominated by fluorescence. This hints to the fact that the peaks shown here refer only to the main backbone

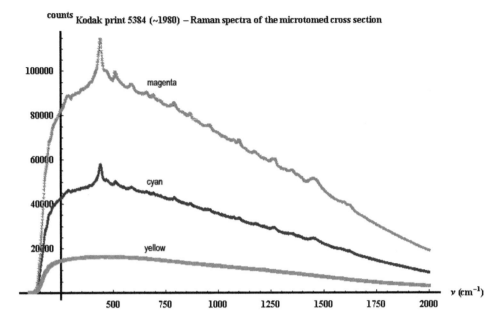

Fig. 24. Raman spectra of each layer of the Kodak print film 5384.

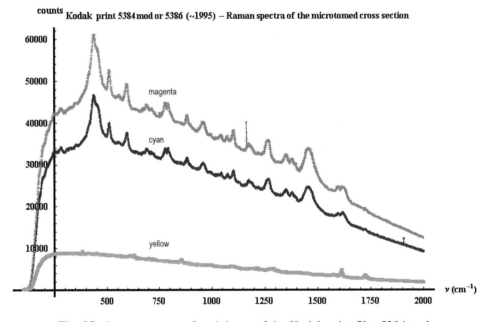

Fig. 25. Raman spectra of each layer of the Kodak print film 5384mod.

of the dyes and not to the characteristic groups. In conclusion, the characteristic Raman peaks are not, even under this geometry, sufficiently intense to be visible, given the background fluorescence.

In the case of the 1970 China girl sample, no characteristic peaks are visible. The question can be raised as to whether or not the large fluorescent signal is due to the age of the film. Fluorescence is expected to increase with ageing, due to the cross-linking of the molecules. To test this, a fresh sample of a Kodak Vision Colour Print stock number 2383 (2004) film was aged in an oven at 70°C and 80% RH. No significant differences between the new and artificially aged film stock were discerned.

3.4.3. Fourier transform infrared spectroscopy

FTIR was used to analyse the compounds separated with TLC. It is important to notice that fresh positive films gave rise to only three lines (yellow, cyan, and magenta), while aged samples gave rise to more lines, usually two lines for each colour, a dominant line and a fainter line. Samples of a recent Kodak print film used for the Colour Dye Fading project running at the National Film and Sound Archive in Australia were dark aged by exposing them to 80% RH and 70°C for a few weeks. The presence of more lines for each colour is an indication that at least one of the ageing process consists of the breaking down of the dye molecule in lower mass compounds.

The lines on the TLC plates were scratched off the plates, dissolved in diethyl ether, and left to decant. The silica particles from the plate were removed, and the coloured solvent was filtered and then dried in a vial under nitrogen. The oily compound left in the vial was mixed with KBr to make pellets suitable for the FTIR analysis at the Research School of Chemistry of the Australian National University.

Figures 26 and 27 show the FTIR spectra for the three dyes of the Kodak print 5384 and the Kodak print 5384 mod (or 5386).

In both cases, the spectra of the three dyes are different, meaning that FTIR can distinguish between the different dyes. Moreover, the spectra of the cyan, magenta, and yellow dyes from the two film stocks have different peaks, indicating that FTIR can identify the dyes present in the film stocks uniquely, and that these film stocks have indeed a different dye formulation.

Ageing the film at 70°C and 85% RH, TLC results in more spectral lines, usually two lines for each colour. It was not possible to analyse the fainter lines because they did not contain enough material. The analysis of the major lines in fresh and aged samples evidenced that differences in the FTIR spectra are present in the magenta line but not in the yellow and cyan lines, as shown in Figs. 28 and 29.

3.5. Conclusions

This work has shown that HPLC is a very promising dye separation technique. This technique implies the destruction of a very small sample of film, and it is therefore not destructive to the film content. The use of confocal Raman micro-spectroscopy failed because of specimen fluorescence and the fact that the dyes are not strongly Raman active. Moreover, it is important to stress that the very basic tool we need, and still do not have,

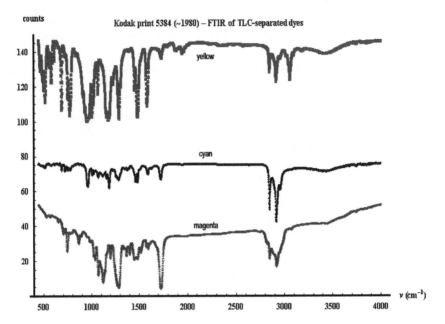

Fig. 26. FTIR spectra for the three dyes of the Kodak print 5384.

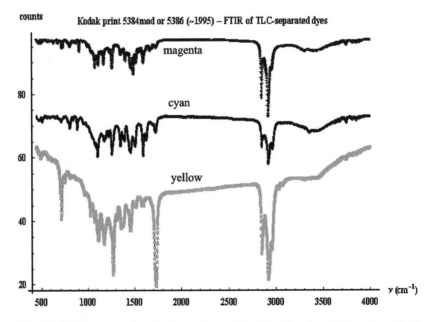

Fig. 27. FTIR spectra for the three dyes of the Kodak print 5384mod or 5386.

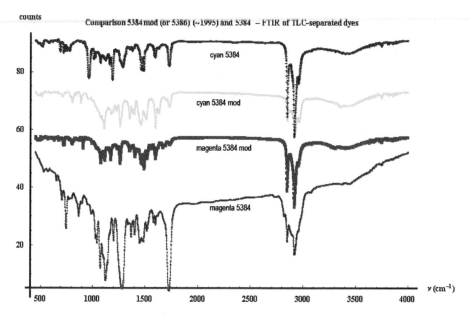

Fig. 28. Comparison between the FTIR spectra of the cyan and magenta dyes of the film stocks 5384 and 5384mod.

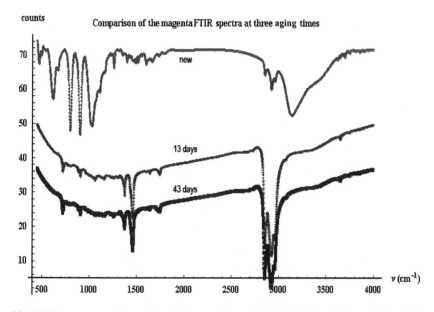

Fig. 29. FTIR spectra of the major magenta line of Kodak Vision Colour Print stock number 2383 fresh and aged at 70°C and 85% RH.

to assess the colour stability of the films stored in an archive, is a clear timeline of the evolution of the film stocks and a working but simple method to identify the film stocks. More work has to be done in this direction and in developing this tool and other simple methods to assess the colour stability of colour motion picture films.

Major questions on silver mirroring have been answered. However, there is the need for more research to understand and assess the permanence of colour photographic materials.

We are in a crucial historical moment for classical photography. The industry is shifting rapidly towards digital photography, and the knowledge acquired on the degradation of photographic materials and stored in the research and development departments of the photographic industries is going to be lost. For conservators in museums and archives, this knowledge is essential because it directly relates to the permanence of their collections. Therefore, it is the right time to invest our energies in this direction.

REFFERENCES

Adelstein, P.Z., Reilly, J.M., Nishimura, D.W., Erbland, C.J., 1992a. Stability of cellulose ester base photographic film: part I – laboratory testing procedures. *SMPTE J.* **101**(5) 336–346.

Adelstein, P.Z., Reilly, J.M., Nishimura, D.W., Erbland, C.J., 1992b. Stability of cellulose ester base photographic film: part II – practical storage considerations. *SMPTE J.* **101**(5) 347–353.

Adelstein, P.Z., Reilly, J.M., Nishimura, D.W., Erbland, C.J. 1995a. Stability of cellulose ester base photographic film: part III – measurement of film degradation. *SMPTE J.* **104**(5), 281–291.

Adelstein, P.Z., Reilly, J.M., Nishimura, D.W., Erbland, C.J., 1995b. Stability of cellulose ester base photographic film: part IV – behavior of nitrate base film. *SMPTE J.* **104**(6), 359–369.

Adelstein, P.Z., Reilly, J.M., Nishimura, D.W., Erbland, C.J., Bigourdan, J.-L. 1995c. Stability of cellulose ester base photographic film: part V – recent findings. *SMPTE J.* **104**(7), 439–447.

Barger, M.S., Hill, T.T., 1988. Thiourea and ammonium thiosulfate treatments for the removal of "silvering" from aged negative materials. *J. Imaging Technol.* **14**(2), 43–46.

Bell, S.E.J., Bourguignon, E.S.O., Dennis, A., 1998. Analysis of luminescent samples using subtracted shifted Raman spectroscopy. *Analyst* **123**, 1729–1734.

Belloni, J., Mostafavi, M., Marignier, J.L., Amblard, J., 1991. Quantum size effects and photographic development. *J. Imaging Sci.* **35**, 68–74.

Bennett, H.E., Peck, R.L., Burge, D.K., Bennett, J.M., 1969. Formation and growth of tarnish on evaporated silver films. *J. Appl. Phys.* **40**(8), 3351–3360.

Bergthaller, P., 2002a. Couplers in colour photography – chemistry and function, part 1. *Imaging Sci. J.* **50**, 153–186.

Bergthaller, P., 2002b. Couplers in colour photography – chemistry and function, part 2. *Imaging Sci. J.* **50**, 187–230.

Bergthaller, P., 2002c. Couplers in colour photography – chemistry and function, part 3. *Imaging Sci. J.*, **50**, 233–276.

Brandt, E.S., 1987. Mechanistic studies of image stability. 3. Oxidation of silver from the vantage point of corrosion theory. *J. Imaging Sci.* **31**(5), 199–207.

Bristow, A.W.T., Courbariaux, Y., Sewell, C., Strawn, A. W., 1998. The isolation and identification of dyes in aqueous media by solid phase extraction-Fourier transform-Raman spectroscopy (SPERS). *Anal. Commun.* **35**, 297–299.

British Journal of Photography 1901. The permanency of toned bromide prints. *Br. J. Photography* **18**, 39.

British Journal of Photography 1918. Degrees of permanence in photographic prints. *Br. J. Photography* **15**, 74–76.

British Journal of Photography 1922. Bloom on negatives and prints. *Br. J. Photography* **25**, 502–503.

British Journal of Photography 1982. The deterioration of gelatin plates. *Br. J. Photography* **12**, 1206.

Curtis, N., Leaist, D.G., 1998. Interdiffusion of aqueous silver nitrate and potassium chromate and the periodic precipitation of silver chromate liesegang bands. *Ber. Bunsenges* **102**(2), 164–176.

Di Pietro, G., 2004. A local microscopic model for the formation of silver mirroring on black and white photographs. In *Proceedings of the International Conference on Metals Conservation*, Canberra, Australia, 4–8 October 2004, pp. 126–136.

Di Pietro, G., 2005. The analysis of dyes in color motion picture films. *Internal Report of the National Screen and Sound Archive in Australia*.

Di Pietro, G., Ligterink, F.J., 2002. Silver mirroring edge patterns. A diffusion-reaction model for the formation of silver mirroring on silver gelatin glass plates. *J. Am. Inst. Conserv.* **41**(2), 111–126.

Di Pietro, G., Mahon, P.J., Creagh, D.C., Newnham, M., 2005. The identification of photographic dyes in cultural materials using Raman spectroscopy. *Proceedings of the 8th International Conference on "Non-Destructive Investigations and Microanalysis for the Diagnostics and Conservation of the Cultural and Environmental Heritage*, Lecce, Italy, 15–19 May 2005.

Eder, J.M., 1903. *Ausführliches Handbuch der Photographie*. Volume 3, *Die Photographie mit Bromsilber-Gelatine und Chlorsilber-Gelatine*. Halle a.S: Knapp.

Eggert, J., 1947. Zur katalytischen Abscheidung von Silber. *Helv. Chim. Acta* **30**, 2114–2119.

Feldman, L.H., 1981. Discoloration of black and white photographic. prints. *J. Appl. Photogr. Eng.* **7**(1), 1–9.

Fleckenstein, L.J., 1977. Color forming agents. In *The Theory of the Photographic Process*. Ed. James, T.H., New York: MacMillan 1977, 4th ed., pp. 339–353.

Fogg, P.G.T., Gennard, W., 1991. *Solubility of Gases in Liquids*. New York: John Wiley & Sons.

Franey, J.P., Kammlott, G.W., Graedel, T.E., 1985. The corrosion of silver by atmospheric sulfurous gases. *Corrosion Sci.* **25**(2), 133–143.

Gillet, M., Garnier, C., Flieder, F., 1986. Glass plate negatives, preservation and restoration. *Restaurator* **7**(2), 49–80.

Graedel, T.E., Franey, J.P., Gualtieri, G.J., Kammlott, G.W., 1985. On the mechanism of silver and copper sulfidation by atmospheric H_2S and OCS. *Corrosion Sci.* **25**(12), 1163–1180.

Grunthaner, F.J., 1987. Fundamentals of X-ray photoemission spectroscopy. *MRS Bull.* **12**(6), 60–64.

Hammond, J.S., Gaarenstroom, S.W., Winograd, N., 1975. X-ray photoelectron spectroscopy studies of cadmium-oxygen and silver-oxygen surfaces. *Anal. Chem.* **47**, 2193–2199.

Hayat, M.A., 2000. *Principles and Techniques of Electron Microscopy*. Cambridge: Cambridge University Press.

Hendriks, K.B., 1984. *The Preservation and Restoration of Photographic Materials in Archives and Libraries: A RAMP Study with Guidelines*. Paris: Records and Archives Management Program, UNESCO.

Hendriks, K.B., 1989. The stability and preservation of recorded images. In *Imaging Processes and Materials*, Neblette's Edition 8. Ed. Sturge, J., Walworth, V., Shepp, A., New York: Van Nostrand Reinhold, pp. 637–684.

Hendriks, K.B., 1991a. On the mechanism of image silver degradation. In *Sauvegarde et conservation des photographies, dessins, imprimés et manuscrits*. Actes des journées internationales d'études de l'ARSAG: Paris, 30 September–4 October 1991, pp. 73–77.

Hendriks, K.B., 1991b. *Fundamentals of Photograph Conservation: A Study Guide*. Toronto: Lugus publications.

Henglein, A., 1993. Physicochemical properties of small particles in solution: microelectrode reactions, chemisorption, composite metal particles and the atom-to-metal transition. *J. Phys. Chem.* **97**(21), 5457–5471.

Henn, R.W., Wiest, D.G., 1963. Microscopic spots in processed microfilms: their nature and prevention. *Photogr. Sci. Eng.* **7**(5), 253–261.

Hermans, J. J. 1947. Diffusion with discontinuous boundary. *J. Colloid Scie.* **2**, 387–398.

James, T.H., 1939. The reduction of silver ions by hydroquinone. *J. Am. Chem. Soc.* **61**, 648–652.

Jahr, R., 1930. *Handbuch der wissenschaftlichen und angewandten Photographie*. Volume 4, *Die Fabrikation der photographischen Trockenplatten*. Wien: Verlag von Julius Springer.

Kejser, U.B., 1995. Examination of photographs with TEM sample preparation and interpretation of the image. In *Research Technique in Photographic Conservation*, Proceedings of the Conference, Copenhagen, 14-19 May 1995. The Royal Danish Academy of Fine Arts: Copenhagen, pp. 41–45.

Kurosaki, T., Itamura, S., Rokutanda, S., Teramoto, T., Miyagawa, T., 1988. Fading of colour photographic dyes III-Isolation and identification of reductive fading products of 1-phenyl-3-methyl-4-aza-(p-diethylaminophenyl)-2-pyrazoline-4,5-dione by bisulphite, *J. Photogr. Sci.* **36**, 79–82.

Lavedrine, B., 1991. Les Aristotypes. In *Les documents graphiques et photographiques*. Paris: Archives Nationales, pp. 149–219.

Levenson, G.I.P., Twist, P.J., 1973. The reduction of silver ions on nuclei by hydroquinone. *J. Photographic Sci.* **21**, 211–219.

Lilienfeld, S., White, C.E., 1930. A study on the reaction between hydrogen sulfide and silver. *J. Am. Chem. Soc.* **52**, 885–892.

McCamy, C.S., Pope, C.I., 1965. Current research on preservation and archival records on silver-gelatin type microfilm in roll form. *J. Res. Nat. Bur. Stand. A. Phys. Chem.* **69A**(5), 385–395.

McCamy, C.S., Wiley, S.R., Speckman, J,A., 1969. A survey of blemishes on processed microfilms. *J. Res. Nat. Bur. Stand. A. Phys. Chem.* **73A**(1), 79–84.

Miyagawa, T., Shirai, Y., 1985. Fading of colour photographic dyes I – reductive fading reactions of 1-phenyl-3-methyl-4-aza-(p-diethylaminophenyl)-2-pyrazoline-4,5-dione and 4-aza-(4'-diethylaminophenyl)-naphtoquinone with thiosulfate. *J. Imaging Sci.* **29**, 216–218.

Moore, W.J., 1972. *Physical Chemistry*. London: Longman.

Nielsen, U.B., 1993. *Silver mirror on photographs*. MS Thesis, Royal Danish Academy of Fine Arts, School of Conservation, Copenhagen.

Nielsen, U.B., Lavedrine, B., 1993. Etude du miroir d'argent sur les photographies. In *Les documents graphiques et photographiques*. Paris: Archives Nationales, pp. 131–143.

Plieth, W.J., 1982. Electrochemical properties of small clusters of metal atoms and their role in surface enhancement Raman scattering. *J. Phys. Chem.* **86**, 3166–3170.

Pope, D., Gibbens, H.R., Moss, R.L., 1968. The tarnishing of Ag at naturally-occurring H_2S and SO_2 levels. *Corrosion Sci.* **8**, 883–887.

Reagor, B.T., Sinclair, J.D., 1981. Tarnishing of silver by sulfur vapor: film characteristics and humidity effects. *J. Electrochem. Soc.* **128**(3), 701–705.

Sakanoue, K., Furutachi, N., 1988. On the dark stability of the magenta azomethine dye image obtained from two-equivalent pyrazolone couplers. *J. Photogr. Sci.*, **36**, 64–67.

Shaw, W.B., 1931a. Permanent bromide prints. *Br. J. Photography* **2**, 594–595.

Shaw, W.B., 1931b. Permanent bromide prints. *Br. J. Photography* **27**, 708–710.

Shuman, D.C., James, T.H., 1971. Kinetics of physical development. *Photogr. Sci. Eng.* **15**, 42–47.

Tanaka, T., Kashiwagi, A., Umehara, M., Tamura, M., 1973. Drift motion of silver ions in gelatin films and its implication in the photolysis of low-pAg emulsions. In *Proceedings of the Symposiums on Photographic Sensitivity*, Gouvill & Caius College and Little Hall, Cambridge, September 1972. R.J. Cox: Cambridge, pp. 139–147.

Theys, R.D., Sosnovsky, G., 1997. Chemistry and color photography. *Chem. Rev.* **97**, 83–132.

Torigoe, M. *et al.* 1984. A challenge in the preservation of black-and-white photographic images. *Sci. Publ. of the Fuji Photo Film Co. Ltd.* **39**(29), 31–36.

Tuite, R.J., 1979. Image stability in color photography, *J. Appl. Photogr. Eng.* **5**(4), 200–207.

Volpe, L., Peterson, P.J., 1989. The atmospheric sulfidation of silver in a tabular corrosion reactor. *Corrosion Sci.* **29**(10), 1179–1196.

Weast, R.C., 1985. *CRC Handbook of Chemistry and Physics*, Edition 66. CRC Press: Boca Raton, Florida.

Weyde, E., 1955. Das Copyrapid-Verfahren der AGFA. Mitteilungen aus den Forschunglaboratorien der Agfa Leverkusen. *Münich Band* **1**, 262–266.

Chapter 5

Hyperspectral Imaging: A New Technique for the Non-Invasive Study of Artworks

Maria Kubik

Senior Paintings Conservator, Art Gallery of Western Australia, Perth Cultural Centre,
Perth 6000, Western Australia
Email: maria.kubik@artgallery.wa.gov.au

Abstract

Hyperspectral imaging generates an accurate digital record for art conservation. This may be used for monitoring change or damage to paintings, as digital documentation resists deterioration better than photographs. Also, digital imaging can assist in the restoration of artwork. It has already been applied to assessing damage from laser cleaning, where computer-aided comparison of before and after cleaning the images have provided superior results. Finally, digital imaging assists in discovering the history of a piece of art by revealing underdrawings and retouchings. Thus, multiple analysis objectives may be achieved simultaneously, which previously required several different instrumental techniques.

A number of applications of this technique are given. In particular, studies of blue pigment in Sydney Nolan and Ivor Hele paintings will be described. The use of the technique to monitor the degradation of varnish on paintings is discussed.

Keywords: Hyperspectral imaging, paintings conservation, pigment identification.

Contents

Physical Techniques in the Study of Art, Archaeology and Cultural Heritage
Edited by D. Creagh and D. Bradley

1. INTRODUCTION

"Colours in the object are nothing but a disposition to reflect this or that sort of rays more copiously than the rest (Rosen *et al.*, 2000)"

Newton's *Optiks*, 1704

Reflectance spectroscopy has already been applied successfully to conservation through the use of UV–Vis–NIR spectrophotometers (Barnes, 1939a; Cordy and Yeh, 1984; Bacci, 2000; Johnston-Feller, 2001; Berns *et al.*, 2002; Oltrogge *et al.*, 2002; Ammar *et al.*, 2003) and Fibre-optic reflectance spectroscopy (FORS) (Bacci *et al.*, 1991; Bacci *et al.*, 1992; Bacci, 1995; Bacci *et al.*, 1998; Picollo and Porcinai, 1999; Bacci, 2000; Bacci *et al.*, 2001; Leona and Winter, 2001; Dupuis *et al.*, 2002). By combining these existing technologies with digital imaging, a new opportunity for pigment identification is made possible. This is achieved by comparison against specially made spectral databases, while the imaging allows distribution of particular pigments to be established (Bacci, 1995; Attas, *et al.*, 2003). Digital imaging, combined with reflectance spectroscopy, provides an extension of conventional conservation photography, and does not require the removal of samples or even contact with the surface (Roselli and Testa, 2005; Torre *et al.*, 2005). Hyperspectral imaging has already been applied in astronomy (Greiner, 2000), remote sensing (Adams *et al.*, 1993; Ichoku and Karnieli, 1996; Prost, 2001), and medical sciences for many years, but only recently has it been extended to the imaging of artwork (Baronti *et al.*, 1997; Baronti *et al.*, 1998; Melessanaki *et al.*, 2001; Casini *et al.*, 2002a; Mansfield *et al.*, 2002; Attas *et al.*, 2003; Berns *et al.*, 2003; Day, 2003a; Roselli and Testa, 2005). Until recently, imagers were restricted to a few broad wavelength bands by the limitations of detector designs and the requirements of data storage, transmission, and processing. Recent advances in these areas have allowed the design of hyperspectral imaging systems that have spectral ranges and resolutions comparable to ground-based spectrometers (Wu and Murray, 2003; Geladi *et al.*, 2004). A convenient application of hyperspectral imaging spectroscopy is realised by means of a camera and a set of narrow-band optical filters. This offers significant cost savings, removes the need for standardised lighting, destructive sampling, and becomes a fast and portable non-sampling technique through instrument design.

2. THE PRINCIPLES OF REFLECTANCE AND HYPERSPECTRAL IMAGING

2.1. Definition of the problem

Whereas conventional reflectance spectroscopy has already found favour in the conservation community due to its non-destructive nature and application to colour analysis, hyperspectral imaging expands the range of possible applications: identifying pigments, monitoring colour changes, colour matching, and the documentation of IR reflectance and induced UV fluorescence (Berns and Imai, 2002; van der Weerd, 2002; Day, 2003a; Pellegri and Schettini, 2003). Early forms of reflectance spectroscopy have been limited, in the first instance, due to large cumbersome instruments with a limited-size sample compartment. Similar to other non-contact techniques, such as fluorescence and Raman spectroscopies, traditional reflectance spectroscopy can only be used for spot testing, whereas art objects have high spatial heterogeneity (Williams *et al.*, 1995; Casini *et al.*, 2002a).

There is a need to develop a cost-effective hyperspectral imaging system that can capture UV-induced fluorescence, visible spectral reflectance, and IR reflectance in specific spectral bands. Much work has already been performed to measure the spectral reflectance of pigments using the visible light range (400–700 nm); however, many pigments begin to show discrimination only into the NIR. The range of wavelengths tested will therefore be increased to match the performance capacity of the Charge-coupled device (CCD) camera to 300–1000 nm. As a reference spectral library for modern pigments is practically non-existent (van der Weerd *et al.*, 2003), it was necessary to focus on collecting spectra from a wide selection of paints and pigments. Therefore, this chapter presents a novel approach to reflectance spectroscopy, identifying pigments in pictorial layers of art, collecting spectra of a large range of artists' pigments in varying binding media, and exploring further opportunities for its application, such as infrared and UV imaging (Mairinger, 2000a).

Investigation of the colour of artists' pigments was undertaken by the Fogg Art Museum at Harvard University as early as 1939 (Barnes, 1939a; Leona and Winter, 2001), where pigments in oil- and water-based binding media were analysed using a recording photo-electric spectrophotometer. Despite its unique characteristics and early popularity, reflectance spectroscopy did not become a dominant analytical tool due to hurdles such as poor wavelength resolution and fingerprinting ability in comparison with other techniques such as FTIR, as well as difficulties in coupling the analysed surface efficiently to the instrument.

Combining digital technology with wavelength-specific filters overcomes the problems associated with traditional reflectance spectroscopic techniques. Known as multispectral or hyperspectral imaging spectroscopy, depending on the number of wavelength bands, multiple channels are used to capture reflectance across the spectral range. From this, a reflectance spectrum may be reconstructed, which allows materials to be distinguished by their composition (Attas *et al.*, 2003). Such a non-contact, *in situ* spectroscopic technique is capable of providing a unique insight into the material composition of the object. Although the extraction of diagnostic information from reflectance and fluorescence spectra is not straightforward, these techniques have the advantage of being non-destructive and easily applicable *in situ*.

The technique is based on irradiating the painting surface with broadband continuous sources, such as halogen lamps, and detecting with a suitable device the back-scattered radiation within narrow spectral intervals. More specifically, each image in the series represents the reflectivity of the imaged object for optical wavelengths within a spectral bandwidth. Data is in the form of a large number of images corresponding to different wavelengths across the visible and NIR part of the spectrum, which in turn can provide a full diffuse reflectance spectrum at each image point (Melessanaki *et al.*, 2001).

By using a high-quality CCD that is sensitive to UV and NIR light, imaging spectroscopy (IS) can be used to retrieve spectral information over the entire wavelength range covered by the camera. Spectral information is obtained with a series of interference filters or imaging monochromators, which operate as a tunable optical filter. A sequence of images is obtained, one for each filter. By stacking together these images from the same area, a multilayer image is formed (Fig. 1). This image data cube consists of as many image layers as channels or filters used, each layer representing an image

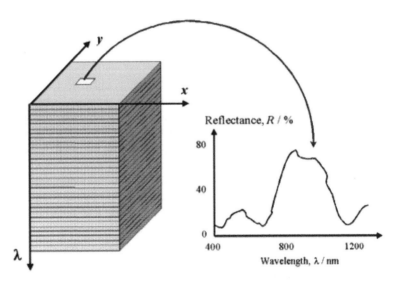

Fig. 1. Stacking multispectral images allows each point to be extracted as a reflectance spectrum (Liew, 2001).

acquired at a particular wavelength band. In hyperspectral images, the resultant dataset is much larger than multivariate images due to the use of more wavelength bands (Geladi *et al.*, 2004). When the number of images increases, their interpretation becomes more difficult (Baronti *et al.*, 1998), so a software program is required to process the resultant data.

2.2. Principles of reflectance

Hyperspectral imaging spectroscopy records the spectral reflection characteristics of a painted surface for individual spectral bands across the UV, visible, and NIR range. The interaction of compounds with radiation across this range are essentially the same. It is governed by a combination of reflection, absorption, scattering, and emission processes, each dependent on the geometry, refractive index, and physical nature of the surface, and wavelength of the radiation used (Turner, 1967; Hapke, 1993; Mairinger, 2000b). In studying the reflected radiance of the illuminated paint layer, it has to be recalled that this layer is composed of coloured pigment particles of various sizes mixed with a binding medium, which can cause enormous departures from homogeneity. The illumination is reflected to some extent from the surface of the paint layer contacted. A smooth surface gives regular reflection, also called specular reflection, in which incident parallel rays remain parallel after reflection. An uneven surface gives diffuse reflection, since the reflected rays are scattered randomly and not in parallel. Some of the radiance is refracted into the paint layer, where it undergoes further reflection by surface facets of larger particles, diffraction (bending of light around larger particles), and scattering (diffusion or deflection of light particles by collisions with small particles suspended in the paint layer). The amount of diffuse scattering is wavelength dependent. Pigment particles selectively absorb certain wavelengths of the radiation, which gives cause to their colour appearance. Multiple reflection, diffraction, scattering, and self-shadowing may occur. The penetration depth of the radiation into the paint layer depends on its wave length, with longer wavelength radiation penetrating to a greater extent, even down to the painting's ground layer in the NIR range. The useful range of wavelength for this research is limited to 400–1000 nm due to varnish interference below 400 nm (See Section 4.2.4). In hyperspectral imaging, the totality of radiation reflected and scattered/diffracted in the direction of the multispectral sensor is recorded for the various sensor bandwidths. For the purpose of this chapter, all non-specular reflection thus recorded is collected under the term "diffuse reflection".

Reflectance spectroscopy is usually concerned with diffuse radiation, and the experimental setup must avoid recording specular reflection. Reflectance varies with wavelength for most materials because energy at certain wavelengths is scattered or absorbed to different degrees. Pronounced upward and downward deflections of the spectral curves mark the wavelength ranges for which the material selectively reflects and absorbs the incident energy. These features are commonly called absorption bands. The overall shape of a spectral curve, and the position and strength of absorption bands in many cases can be used to identify and discriminate different chemical compounds. Spectral features in paints tend to be broad (Best *et al.*, 1995). Wavelength-specific absorption in pigments is caused by the presence of particular chemical elements or ions, the ionic charge of certain elements, the geometry of chemical bonds between elements, and the presence of transition elements in their crystal structures (Burns, 1993).

2.3. Kubelka–Munk theory

A variety of reflection models have been developed to address paint layer behaviour under illumination. The critical model component is on light reflection and scattering within the paint layer. Kajiya (1985) proposed to derive the reflection based on the Kirchhoff theory, which calculates an electromagnetic field over the paint layer. This method was further developed by He *et al.* (1991) where many factors, including layer statistics and subsurface scattering, are considered. Another class of techniques have been proposed by Kubelka and Munk (Kubelka and Munk, 1931; Turner, 1967; Curiel *et al.*, 2002) to study diffuse reflectance of dispersed powders, deriving a highly non-linear relationship between pigment concentration and spectral reflectance. As early as 1949, the Kubelka–Munk transformation (KMT) was shown to be a reliable model for the mixing behaviour of paint mixtures, which has since been applied by the paint industry. KMT can be extended to model realistic situations to provide a clear and powerful tool for evaluation of some of the optical properties of particulate thick films (Harrick Scientific, 2006). In principle, the theory reduces to a spectral absorption and scattering ratio, which is nearly unique for a given pigment (Morgans, 1982; Wendland and Hecht, 1996). The Kubelka–Munk (K–M) model has a particularly simple solution in the case of semi-infinite samples. All the geometric peculiarities of the inhomogeneous sample are condensed into a single parameter, the scattering coefficient s. The diffuse reflectance R_∞ is given as (Harrick Scientific, 2006):

$$R_\infty = 1 + \frac{k}{s} - \sqrt{\frac{k}{s}\left(2 + \frac{k}{s}\right)}$$

where k is the absorption coefficient of the sample ($k = 4\pi\kappa/\lambda$ where λ is wavelength). This relatively simple form is easily solved for k/s, yielding the familiar Kubelka–Munk transform:

$$\frac{k}{s} = \frac{\left(1 - R_\infty\right)^2}{2R_\infty}$$

The KMT of the measured spectroscopic observable R_∞ is approximately proportional to the absorption coefficient, and hence is approximately proportional to the concentration. The scattering coefficient was introduced into the theoretical description of diffuse reflection as a semi-empirical parameter to account for internal scattering processes. The scattering coefficient s is, in fact, dominated by particle size and refractive index of the sample (Turner, 1967). It is not a strong function of the wavelength or the absorption coefficient, so the KMT model considers it to be a constant. In reality, the scattering coefficient varies slowly with wavelength (Berns *et al.*, 2002). More importantly, it changes significantly with packing density, so care should be taken to pack powdered samples as reproducibly as possible if quantitative results are desired. Due to its simplicity, the latter expression for *k/s* has been widely incorporated as the diffuse reflectance transform in the standard infrared spectroscopy software of commercial FT-IR spectrometers. As *k/s* values are assumed to be additive, the KMT is particularly useful in mixture analysis.

2.4. Opportunities for hyperspectral imaging spectroscopy

Besides identifying pigments, hyperspectral imaging generates an accurate digital record for art conservation. This may be used for monitoring change or damage to paintings, as digital documentation resists deterioration better than photographs (Casini *et al.*, 2002a). Digital imaging may also assist in the restoration of artwork. It has already been applied to assessing damage from laser cleaning (Balas *et al.*, 2003), where computer-aided comparison of before and after images have provided superior results. Finally, digital imaging assists in discovering the history of a piece of art through revealing underdrawings and retouchings (Day, 2003a). Thus, multiple analysis objectives may be achieved simultaneously, which previously had required several different instrumental techniques.

2.4.1. Identification for retouching

Damaged paintings frequently require retouching to provide visual compensation for losses in the paint film, but using the correct pigment for retouching damaged artworks is complicated by the possibility of metamerism. This occurs when two colours that appear similar under one light source look different under another. To make losses indistinguishable from the surrounding undamaged areas, identical reflectance properties in both retouching paint and the original must therefore be achieved. Colour matching, therefore, involves mixing pigments to match a given standard, assuming normal colour vision and a standard source of illumination (Day, 2003a). A complete spectral match is obtained when the same colour is perceived in all light temperatures, which is only achieved when either the same pigments or those with the same reflectance pattern are used (Morgans, 1982; Imai *et al.*, 2000). In this manner, pigment identification plays an important role in conservation.

2.4.2. UV imaging

Photographic documentation of fluorescence emitted under UV exposure has been an important diagnostic method for studying historical and artistic objects since the 1920s (Casini *et al.*, 2002b; Pelagotti *et al.*, 2005). This simple diagnostic method has the capacity to reveal information about an object that is otherwise not visible, such as overpainting and varnishes (Eastman Kodak, 1972). Since many inorganic and organic substances exhibit a characteristic emitted fluorescence under UV illumination, fluorescence spectroscopy theoretically can also be applied to differentiate materials such as resins and pigments.

2.4.3. IR imaging

Similar to UV, near-IR photography has been used in analysing works of art for many decades (Mansfield *et al.*, 2002). Some inks and pigments that are visually identical are frequently different under NIR (780–2500 nm). Near-IR has less energy than visible light and usually excites vibrational overtones rather than electronic transitions (Mairinger, 2000b). The ability of NIR to penetrate through some pigments has allowed the study of underdrawings. The technique of using NIR photography for detection of underdrawings was already in use by the late 1930s, when Ian Rawlins used an NIR camera to improve visual assessment of paintings (Roselli and Testa, 2005). This is due to the low absorption by some pigments in the NIR range (Gargano *et al.*, 2005), and it is thus possible to divide between carbon-based (*e.g.* carbon black), iron-oxide-based (Mars Black), and organic-based compounds (sepia, bistre, and iron gall ink) (Mairinger, 2000b; Attas *et al.*, 2003), as is increasing the legibility of texts obscured by dirt, deterioration, bleaching, or mechanical erasure. Generally, a paint layer becomes more transparent with greater wavelength of the incident radiation, smaller thickness of the paint layer, smaller number of particles in the layer, and lesser refractive index difference between pigment and medium (Mairinger, 2000b). Some work has already been performed with the aid of CCD technology; some pigments were found to become transparent between 800–1100, others in all the NIR region, and others only beyond 1000 nm (Gargano *et al.*, 2005). Maximum transmittance for many pigments occurs between 1800 and 2200 nm, while >2000 nm is required to penetrate blue and green layers (Mairinger, 2000b). The NIR region is also applied for materials identification (Attas *et al.*, 2003). Mansfield *et al.* were able to differentiate between pigments using digital images taken between 650 and 1100 nm, finding sufficient specificity in this region (Mansfield *et al.*, 2002). Clarke also chose to use NIR imaging in favour of UV–Vis spectroscopy for pigment identification. He concluded that imaging between 700 and 950 nm was useful for certain pigments, particularly blues found on early mediaeval manuscripts (Clarke, 2004). The entire visible and NIR region cannot be covered by a single camera (Gargano *et al.*, 2005), so a combination of two systems is recommended. Gargano *et al.* (2005) found that a good Si CCD camera facilitated adequate detection work up to 1000 nm. For infrared imaging beyond this wavelength, NIR-sensitive diode arrays are required, such as an InGaAs (Geladi *et al.*, 2004), PbS (Baronti *et al.*, 1997), or PtSi camera (Gargano *et al.*, 2005).

2.4.4. Colorimetry

"Colour is a subject that ought to give intellectuals a headache. Its definition is so completely intangible".
Sidney Nolan, writing to *Sunday Reed*, April 6, 1943 (Nolan, 1943a).

Measuring the spectral reflectance of an object surface objectively provides a description of the surface's inherent physical characteristics, while colour perception depends on many factors, such as the illumination, the observer, and the surrounding conditions (Barnes, 1939b; Nassau, 2001; Zhao *et al.*, 2004). Colour is determined only by absorbance in the visible spectrum, and conversely, colour assignment can, easily be calculated from spectral reflectance. The three subjective natures of colour are hue (wavelength), saturation (purity), and luminosity (intensity of reflected light). These primaries are assigned the tristimulus values X, Y, and Z, respectively, obtained by measurement of the object reflectance and source emission spectra, and may be used to quantitatively describe colour according to Commission Internationale de l'Eclairage (CIE) conventions (McLaren, 1983). CIE Y value is a measure of the perceived luminosity of the light source, whereas X and Z components give the colour or chromaticity of the spectrum. Currently, the CIELAB 1976 system is the most widely used for describing colour changes. This definition starts from the tristimulus values defined, and defines three other quantities, $L*$, $a*$, and $b*$, where:

$L*$ = Lightness (0 = black, 100 = white)

$a*$ = Magenta–Green (−128 = green, +127 = magenta)

$b*$ = Yellow–Blue (−128 = blue, +127 = yellow)

CIE $L*a*b*$ measurements provide a permanent colour record (Morgans, 1982) and are calculated to linearise the perceptibility of colour differences. Each colour can thus be represented in a 2D or 3D chromaticity diagram that depends on the adopted CIE convention. In the 2D diagram, the area on the graph surrounded by the triangle is known as the spectrum locus and covers the visible wavelengths 400–700 nm (Fig. 2).

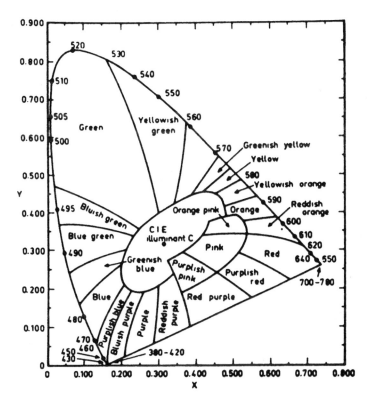

Fig. 2. Two-dimensional CIE Lab diagram (Frei and MacNeil, 1973) allowing the plotting of colour in functions of X, Y, and wavelength.

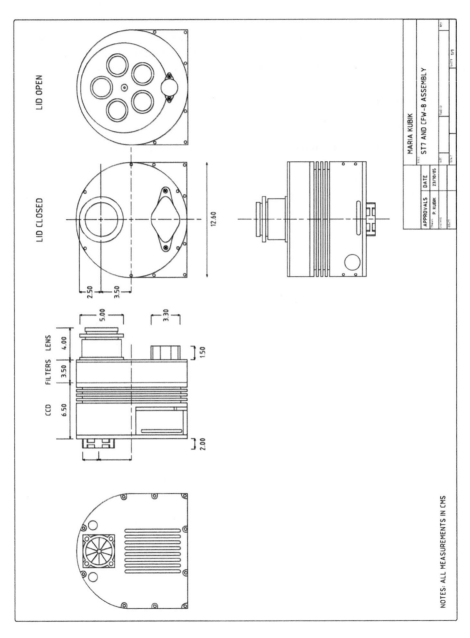

Fig. 3. Technical drawing of CCD setup with filter carrousel and lens.

3. BUILDING A HYPERSPECTRAL IMAGING SYSTEM

Much of the methodology for this project involved establishing and building a system suitable for the required tasks. The design principle was to create a simple and cost-effective system using a high-resolution CCD detector with a wide wavelength response, integrated with a number of optical filters ranging between IR and UV bands (Fig. 3). This was combined with a laptop running dedicated software to steer instruments, record, and manipulate data. Reflected light intensity from a broad range source may then be recorded as a function of wavelength and location. To be effective, the system must be robust and portable, it should be fast in setup and collection times, and must be built from readily obtainable components to minimise costs.

3.1. Lighting

The excitation source must have even, broad-range emission, and should be bright enough that the exposure time is relatively short. A number of lighting systems have already been applied in previous research, most commonly consisting of quartz tungsten halogen (QTH) lights found in slide projector lamps (Greiner, 2000; Melessanaki *et al.*, 2001; Attas *et al.*, 2003). These are rated between 100–300 W, and emit radiation from 380 to well beyond 1050 nm (Casini *et al.*, 2002a). The spectral output of such lamps is fairly consistent, leading to low variability in the recorded reflectance. A spectral curve may be seen in the information provided by Oriel Instruments (www.oriel.com) for their 6315 QTH lamp (Fig. 4).

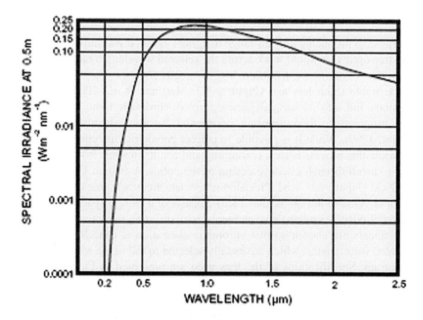

Fig. 4. Typical emission diagram for a quartz tungsten halogen lamp (Oriel Instruments).

A second option was proposed by Saunders and Cupit (Saunders and Cupitt, 1993). Light output was projected through filters and then fed through optical fibres, rather than introducing the filter array at the collection end, providing more even distribution of illumination. This was, however, not pursued since optical fibres do not illuminate a large surface area, the detector camera would be sensitive to ambient light, and two sets of filters would be required for the two illumination sources, doubling the cost of the most expensive system component.

Pursuing the QTH option, two projectors are placed at 45° angles with respect to the normal to the surface to provide even illumination and reduce glare. The goal was to replicate the setup commonly used when photographing paintings, which provides spatially uniform diffuse illumination.

Based on the spectral response of a calibration standard, *e.g.* a Spectralon plate, source emission may also be described by the measured reflectance across the wavelength region of interest (Fig. 18). While this incorporates the effects of variable filter efficiency, it shows the need for a method of calibration to equalise the response rate. Similarly, while power for the light projector and camera was not drawn from a voltage-stabilised supply, fluctuations could be overcome at the calibration stage.

Lighting can cause glare (specular reflection) on paintings, especially if they are varnished or glazed (Day, 2003a). Any glazing must therefore be removed first. The National Gallery, London, used polarisers to reduce glare during photographing (Day, 2003a). This, however, causes dark areas to appear richer, and the polarising film was also found to be very temperature sensitive, such that it could not be placed over the projectors after the infrared cut-off filters were removed.

3.2. Optics

For the purpose of producing a broad-range imaging system, it was important to choose an aberration-free optic that would work across the selected wavelength range. A high-quality lens, giving rise to little or no distortion, low chromatic aberration, and high resolution, is therefore desirable (Saunders and Cupitt, 1993; Martinez *et al.*, 2002). Aberrations of optical systems fall into various categories. In monochromatic light, these are spherical aberration, astigmatism, field curvature, coma, and distortion (Hardy and Perrin, 1932; Melles Griot, 1999). While it is possible to correct geometric distortions in an image by using software, this process is time consuming and results in some loss of image quality; it is better to avoid through effective design of the optics. A 50-mm UV achromatic lens by Sigma Koki (Japan) was used. Precision achromats are nearly free of spherical aberration and coma. As seen in Fig. 5, such a lens consists of a closely spaced, often cemented combination of positive and negative elements, with differing refractive indices (RI).

These elements are chosen so that chromatic aberration is cancelled at two distinct, well-separated wavelengths, which are usually selected to fall in the red and blue portions of the spectrum. Specifications of the lens used are presented in Figs. 6 and 7, which show the thickness of the individual lens layers (d), the focal length (f), and diameter (D) of the lens.

To test for any distortion, the UV achromat lens was evaluated by acquiring images of a ruled grid and Gamblin squares at different wavelengths throughout the region

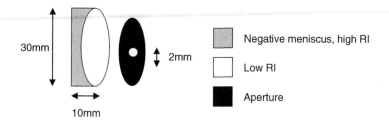

Fig. 5. 50-mm Achromat lens and 2-mm aperture.

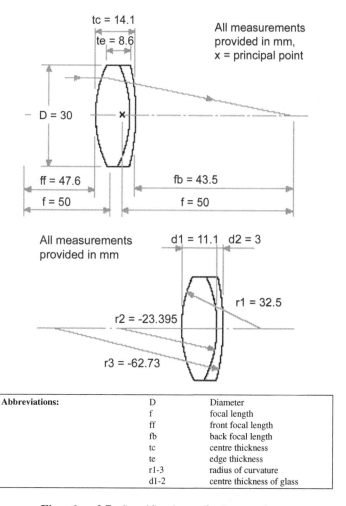

Abbreviations:			
	D		Diameter
	f		focal length
	ff		front focal length
	fb		back focal length
	tc		centre thickness
	te		edge thickness
	r1-3		radius of curvature
	d1-2		centre thickness of glass

Figs. 6 and 7. Specifications of achromat lens.

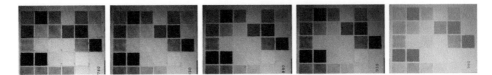

Fig. 8. Five images of Gamblin paints between 780 and 980 nm; these were overlaid to test for geometric distortion.

of interest. Trials showed no distortions. This was seen both in the creation of 3D data cubes of reference grids, and by the correct registration of target samples between 400 and 1000 nm. This is seen, for example, in five images of Gamblin reference paints at wavelengths 780, 830, 880, 930, and 980 nm, which overlap and may be used to form a data cube (Fig. 8). Testing for lens distortion was achieved using a 1-cm² grid placed on the easel. By capturing images at all wavelengths, the possible change in measurements was monitored. Images were captured individually, and the length and width were measured by counting the number of squares at the centre and two sides (both vertical and horizontal direction), so that a total of five measurements were made per image. At a distance of 150 cm the grid measured 27.6 cm height × 41.4 cm width at all wavelengths, while diagonal axes too remained constant. This is confirmed by the field-of-view (FOV) calculations below. Further testing in the range 380–1000 nm showed no change in measurements, nor was a change in focus required due to large depth of field, which kept the image size constant for later overlaying. There are, therefore, no difficulties associated with spherical aberration using this system. For the 9.2 × 13.8 mm CCD chip used, the FOV is calculated as:

$$
\begin{aligned}
\text{FOV} &= \left(\frac{\text{CCD chip size}}{\text{focal length}} \right) \times \text{distance to object} \\
&= \left(\frac{9.2 \text{ mm height}}{50 \text{ mm}} \right) \times 150 \text{ cm} \\
&= 27.6 \text{ cm height} \\
&= \left(\frac{13.8 \text{ mm width}}{50 \text{ mm}} \right) \times 150 \text{ cm} \\
&= 41.4 \text{ cm width}
\end{aligned}
$$

Hence, that the field of view at 150 cm object distance is 27.6 × 41.4 cm. As astronomical CCD cameras, such as the ST-XME8 used, are extremely light sensitive for intended nighttime use, an aperture had to be designed to reduce light throughput and thus detector overload. Apertures of 2 mm diameter and larger were tested depending on the light, and best results were achieved by placing this aperture in front of the lens (Fig. 5). The *f*/stop created by the aperture may be calculated as the ratio between the diameter of the aperture and the focal length of the lens, so that the calculated *f*/stop for the current setup is:

$$\frac{50 \text{ mm}}{2 \text{ mm}} = f25$$

This can next be used to calculate the depth of field (DOF), the range of acceptable focus in the object space, using the following calculation:

$$DOF = \left(\frac{2 \times WD^2 \times F\# \times C}{F^2} \right)$$

Where WD = working distance from the lens to the object (1500 mm), $F\#$ = *f*/stop (*f*/25), C = circle of confusion, measured by pixel size in the sensor plane (9 μm or 0.009 mm), F = focal length of the lens (50 mm), so that:

$$DOF = \left(\frac{2(1500)^2 \times 25 \times 0.009}{50^2} \right) = 405 \text{ mm}$$

Thus, the DOF is 405 mm or 0.405 m, which translates to a workable focal range between 1.005 and 1.905 m when the object distance is 1.5 m from the lens, providing greater flexibility and repeatability in system setup. The lens and aperture are housed in a specially machined aluminium flange with double thread for more accurate focus control.

To eliminate all chromatic aberrations, reflectance optics from a microscope were also considered. A rich-field (or wide-field) reflector is a type of Newtonian reflector with short focal ratios and low magnification. Collection optics consisted of front-surfaced spherical mirrors with working entrance of *f*/2.4 (Carcagni *et al.*, 2005; Roselli and Testa, 2005). This microscope objective was ultimately not pursued for the current project, as there was some vignetting of the image, and they did not provide measurable benefit in comparison to the apochromat lens used.

3.3. CCD array

The array detector is at the centre of the system, and choosing the right system (CCD, CMOS, or diode array) for imaging spectroscopy is crucial. As the system was designed for field use, a small portable CCD, controllable by laptop and thermostatically cooled, was preferable. Key criteria for such a detector included real-time spectral imaging for inspection and focussing, high spectral resolution and range, low dark current, high throughput and sensitivity in order to avoid long exposure times, and intense light excitation, which are otherwise potentially harmful to the object (Balas *et al.*, 2003). The CCD must be able to interface smoothly with other equipment, such as filter wheels and computer technology (Guntupalli and Grant, 2004). It was decided that a complementary metal oxide semiconductor (CMOS) chip would not be used as a cheaper alternative, as at the time of this writing, these detectors have not yet supplanted CCD technology, and CCD camera systems have already addressed the key requirements of scientific imaging and spectroscopy (Guntupalli and Grant, 2004). Likewise, a diode array detector was not selected for the current application as these detectors are not commonly available on the market. Most consumer digital cameras have been determined not to have the required criteria (Day, 2003a), the difference being inadequate size of the CCD chip, freedom from defects, and the amount of cooling provided to the chip.

The CCD has been successfully applied to scientific imaging and spectroscopy, thanks to its outstanding ability to detect and quantify light (Guntupalli and Grant, 2004). The structure and basic applications of the CCD were invented at the AT&T Bell Lab in 1969 (Hamasaki and Ochi, 1996). Improvements in technology have seen an increase in quantum efficiency, decrease in dark current and pixel size, and reduced cost. The performance of CCD technology has been further improved through back illumination, deep depletion for near-IR sensitivity, and on-chip multiplication gain for single-photon sensitivity at high speeds (Guntupalli and Grant, 2004).

In an idealised form, a CCD chip consists of a pixel array that produces electrical charges proportional to the amount of light received at the pixels. The imaging sensor CCD is used both to collect photo-generated charge and to transport it (Fig. 9). The image is captured as photons, which fall onto a semiconductor substrate, the 2D array of pixels (Hamasaki and Ochi, 1996). In the STMEI camera used (Fig. 13), each pixel measures 9×9 μm, and is arranged in a 1020×1530 pixel grid. Each pixel has an intensity value and a location address. The intensity value represents the measured physical quantity, such as reflectance at a given wavelength. Sensitivity is determined by the quantum efficiency (or the fraction of incident photons converted to electrons in the wells of the CCD) and by

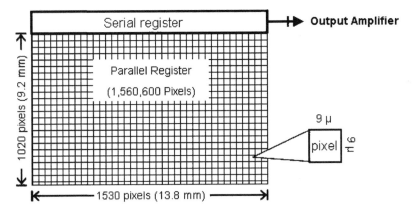

Fig. 9. Specifications of the Kodak KAF 1602E CCD chip incorporated into the imaging system.

the system read noise (Guntupalli and Grant, 2004). Noise may appear on the image and is caused by a variety of factors, as outlined below.

- **Linear pattern noise**, called smear or blooming, occurs due to leakage current and light across pixels and from pixels to serial registers (Hamasaki and Ochi, 1996). This is seen as either vertical white lines on the display under high light conditions or bright white points, and may be eliminated by using overflow drains and anti-blooming devices.
- **Read noise** originates from the onboard pre-amplifier and is dependent on the amount of signal the device is designed to handle (full-well capacity) and the readout speed. On-chip multiplication gain has the advantages of multiplying the charge right on the CCD, before it reaches the preamplifier.
- **Dark current** is the thermally generated charge in the CCD pixels, and can be a significant noise source, especially in experiments requiring long exposure times or in those in which binning is used. This can be reduced by cooling; dark current is halved with every drop in temperature of 5–7°C (Hamasaki and Ochi, 1996; Guntupalli and Grant, 2004).
- **Shot noise** is inherently present in both the dark charge and signal charge. It arises from the quantisation of charge into units of single electrons and for a given number of electrons in a charge packet. Its effects are reduced by using a greater well depth and by reducing the readout dark charge.
- **Non-uniformity noise** results from each pixel producing a slightly different amount of dark current. Unlike shot noise, non-uniformity noise is eliminated by subtracting a dark current frame from each image.
- Other sources of noise include **thermal, $1/f$, photon, and fixed-pattern noise** (Day, 2003a).

Among the most popular CCDs used by amateur astronomers are those made by Santa Barbara Instrument Group (SBIG) (Greiner, 2000). Almost all use Kodak E-type chips, with quantum efficiency (QE) of around 60% between 550 and 750 nm (Fig. 10) (Imai et al., 2003). Both overall sensitivity increase and a more uniform response over the spectrum shortens the exposure times required (Imai et al., 2003). These cameras also have as standard a single-stage thermo-electric cooler. For the 0400 and 1600 series chips, which have 9-μm^2 pixels, the dark current is about 1 electron (e$^-$) per second per pixel at −10°C. Based on the available information, the ST-8XME camera by SBIG was selected. Specifications provided for the camera were:

–CCD Kodak KAF-1602E + TC-237
–Pixel array 1530 × 1020 pixels, 13.8 mm × 9.2 mm
–Pixel size 9 mm × 9 μm
–Full-well capacity (anti-blooming gate) ~50 000 e$^-$
–Dark current 1e$^-$/pixel/s at 0°C
–Electromechanical shutter, exposure time = 0.12–3600 s

The CCD temperature can be controlled to −25°C, although it is recommended that the camera be operated at 75–85% capacity to allow for fluctuating ambient temperature (SBIG, 2003).

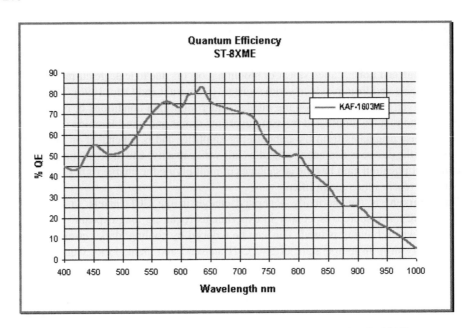

Fig. 10. Spectral resolution of the ST-8XME camera (SBIG, 2003).

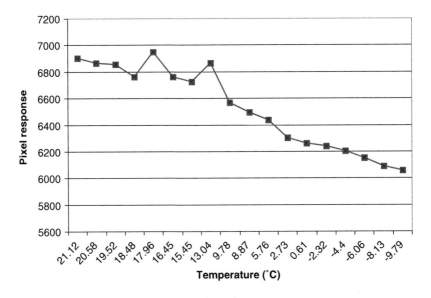

Fig. 11. Interdependence of pixel response to temperature.

This was the equivalent of $-10°C$ in a lab that had an approximate steady temperature of 16°C. Whenever cooling is active and the camera is disconnected or switched off, the "shutdown" option must be used in order to ramp down cooling and prevent the formation of condensation. Cooling the camera is shown to have an effect on reflectance intensity readings (Fig. 11); therefore, it is important to let the camera reach and settle on the desired temperature setting before use. To demonstrate the importance of steady cooling, intensity readings were collected three times at different temperatures from the Spectralon plate at 900 nm. The cooler the camera, the lower the intensity readings became, but the better the residual noise is minimised.

Resolution of the object is governed by the pixel size of the CCD and the optical characteristics of the camera, such as its airy disk size (Vallee and Soares, 2004). The overall resolution of the system was determined by imaging a 1951 USAF calibration plate at the intended object distance of 150 cm (Fig. 12). The limits of resolution can be found at the smallest line detail, where the three vertical lines can just be distinguished. This occurs in row 0, line 6, which corresponds to a resolution (line thickness) of 0.2 mm, and agrees with the CCD pixel size in the object plane. It demonstrates that our optics design did not contribute to resolution losses. In fact, the pixels can clearly be seen to limit the resolution in the enlarged section of the USAF target. Increasing the distance reduces the resolution accordingly.

3.4. Filters

Various filtering technologies are available that provide useful bandwidths for hyperspectral imaging. Filtering can be applied at the camera or at the light source, and the second approach reduces the amount of time an object is exposed to potentially harmful radiation.

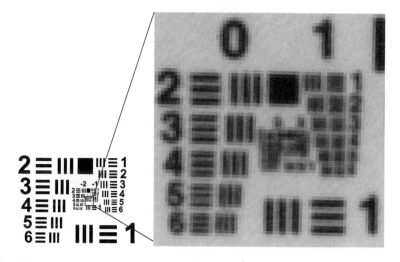

Fig. 12. 1951 USAF target, and smallest resolved line series achieved at column 0 row 6.

While wide-band optical filters (Pellegri and Schettini, 2003) and monochromators (Balas *et al.*, 2003) have been trialled, the most robust systems are those that are able to separate incoming radiation into narrow spectral bands using interference or tunable filters (Eastman Kodak, 1970; Rosen *et al.*, 2002). An interference filter permits selective wavelengths to pass through layers of varying thickness in an optical coating. With an appropriate structure, optical interference processes lead to the transmission of one specified wavelength and the reflection of all others. A liquid crystal tunable filter (LCTF) is different in its ability to change its transmission wavelength electronically. Such a tunable filter has no movable parts, increases the number of available wavelength channels, and is faster compared to traditional filters that must be changed one at a time (Slawson and Schettini 1999; Melessanaki *et al.*, 2001; Imai *et al.*, 2002; Attas *et al.*, 2003; Pellegri and Schettini, 2003). Transmittance of an LCTF, however, is much lower than interference filters, varying from 31% at 720 nm to 5% at 435 nm (Slawson *et al.*, 1999; Imai *et al.*, 2000). Furthermore, there is no single LCTF that covers the spectral region of interest. Therefore, it requires two separate and expensive units, the total cost being almost six times as much as a comprehensive set of interference filters.

The most cost-effective method involves selecting a series of interference or band-pass filters, the number of which determines the cost and range (Baronti *et al.*, 1997; Bell *et al.*, 2002; Martinez *et al.*, 2002; Zhao *et al.*, 2004). A set of narrow-band filters was chosen to span the required spectrum. In this case, 18 filters (across the range 400–1000 nm) were used to coincide with the sensitive range of the CCD. Each filter has an active diameter of 2.0 cm and is lodged in a five-slot filter carrousel, using locking screws. A total of four such carrousel wheels were prepared and could easily be changed between image collections. Carrousels are placed into the filter wheel mount, which is rotated by a stepper motor and activated by computer. This filter wheel is mounted between the painting and the camera lens (Figs. 3 and 13). A micro-switch, which senses the position of notches cut in

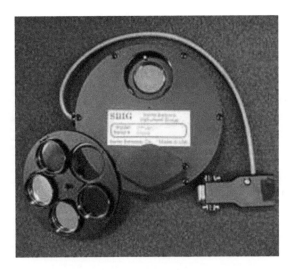

Fig. 13. SBIG STMEI 8 filter wheel and carrousel (SBIG, 2003).

the edge of the wheel, allows the user (via the positioning system interface) to select a specific filter. A rubberised rim provides the traction for turning. As the interference filters used are relatively thick, even a small change in the alignment of the filter produces a significant deviation in the light path (Saunders and Cupitt, 1993), resulting in a displacement between colour separation images of the order of several pixels. This was minimised by the tight-fitting locking screws designed, and by ensuring that the filter wheel was adequately secured before locking the camera into position.

The transmittance of a given filter sample is defined as the ratio of the amount of radiation transmitted by a filter to the amount of radiation incident on it, usually expressed in percentage (Eastman Kodak, 1970). The spectral transmittance data of our filters was provided by the manufacturers (Fig. 14), and their variable percentage transmission shows the need for standardisation through calibration. This chart shows the 22 filters considered; however, only those 18 over 400 nm were used for pigment analysis. Filters at 400, 450, and >780 nm, in particular, have transmission around 50% or less, which must be overcome with extended exposure times. A calibration transform was applied to correct for these variations.

3.5. Image processing

To complete the system, software must be created to drive the various components, capture and manipulate images, and interpret the data obtained. The following processing functions were undertaken.

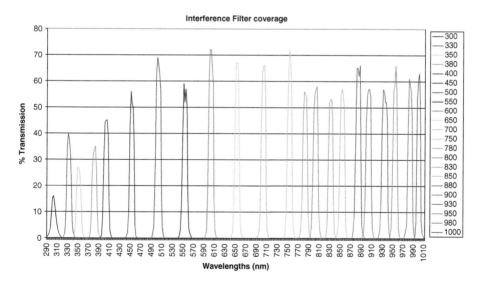

Fig. 14. Interference filter transmission curve.

- *Image capture*: The highest resolution and largest bit depth possible must be captured and saved. Raw data is initially saved and stored in an uncompressed format, such as in the 16-bit TIFF format (Rosen *et al.*, 2002; Day, 2003a). This task is completed using CCDOps (5.39) from SBIG.
- *Data cube creation*: Images are transferred into Hypercube (8.2) as TIFF files with numbered extensions. This hyperspectral imaging program facilitates the creation of a 3D spectroscopic imaging data cube, where X and Y dimensions represent the pixel location of individual images, and Z the wavelength at which each image was taken.
- *Spectral processing*: Transformation from digital signals to reflectance is a critical part of spectral processing (Berns *et al.*, 2002; Imai *et al.*, 2002; Attas *et al.*, 2003). By slicing through any picture or pixel cluster location in the data cube, the reflectance spectrum corresponding to that particular point on the sample may be extracted. If not directly compared to a reference paint captured alongside the target painting, the spectra must be converted to relative reflectance by taking the pixel-by-pixel ratio against the 99% Spectralon reference plate captured alongside the art object. Averaging over several points greatly improves signal-to-noise ratio (van der Weerd *et al.*, 2003), so that for every spectrum, three points were chosen for averaging.
- *Spectral matching*: This is achieved using either Hypercube or ThermoGalactic Grams AI. Similarity between reference and unknown region spectra can be judged by the relative closeness of these positions (spectral distance) or by how small the angle is between spectral vectors.

3.6. Application methodology

While constructing a hyperspectral imaging system formed the overarching component in the methodology, its usefulness was only able to be achieved by building up a set of reference spectra that could either be included in the image, or saved in a separate library of reflectance spectra. In the second approach, samples of paint materials and their spectra make up a calibration or training set for the experimental procedure, and make the identification of unknown materials possible (Balas *et al.*, 2003).

3.6.1. Reference paint samples

As identification of pigments is achieved by comparing spectral reflectances, it is important to have a good reference set of reflectance spectra from a variety of pigments in appropriate media and substrates. In particular, the visual light reflection spectra of most modern pigments are relatively specific (van der Weerd *et al.*, 2003), so that effort will be made to include these pigments in any potential spectral database.

The paint swatches must be tested in an upright manner, as required for the UV/Vis/NIR Spectrophotometer sample compartment, or on an easel as required for the imaging system. As such, dry powders could not be measured, and all pigments had to be mulled into paint format to remain stable in a vertical position. Pigments were hand mulled into linseed oil, gum arabic, and egg yolk to represent different binding media. To test whether the binding agent caused discrepancy in reflectance results, spectra were collected from pigments prepared into

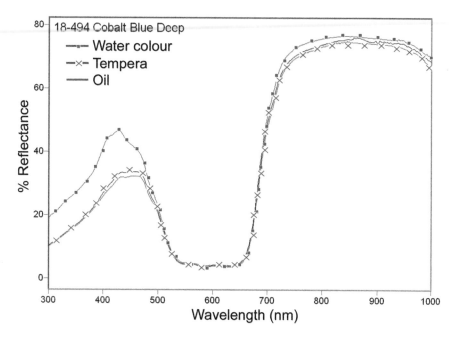

Fig. 15. The effect of binding medium on the reflectance curve – Schmincke cobalt blue deep.

three paints using the binding media of oil, tempera and gum arabic. These spectra could then be superimposed for comparison (Figs. 15 and 16). The basic reflectance curve, with absorbance and reflectance features, remained constant across the three binding media tested.

In all cases, pigment ground in watercolour (gum arabic) showed a higher percentage reflectance as it has a higher chroma than the other binding media. This may also be due to the thinner nature of watercolour paint, which takes advantage of the white ground layer for increased brightness. Pigments ground in oil showed the darkest colour with lowest percentage reflectance curve, while tempera tended to closely resemble the oil paint. The possible effects from fading and/or yellowing on the identification of pigments is discussed later. This also has consequences for paint ageing.

3.6.2. Positioning the painting

The DOF, the final image size, and resolution that determine the working distance between camera and painting are important. This furthermore impacts on the total amount of storage space required, and in the case of large paintings that do not fit into the field of view in one section, the total number of frames to be mosaiced (Baronti *et al.*, 1998; Attas *et al.*, 2003; Day, 2003a). The painting is placed on an easel in front of the camera, and the two projection lights are placed at 45° incident to the surface (Fig. 17).

Any glass or Perspex covers are removed from the object under investigation to minimise glare. Working in a darkroom is not necessary, since the effects of ambient light are cancelled through calibration. The painting, however, must be vertical to prevent

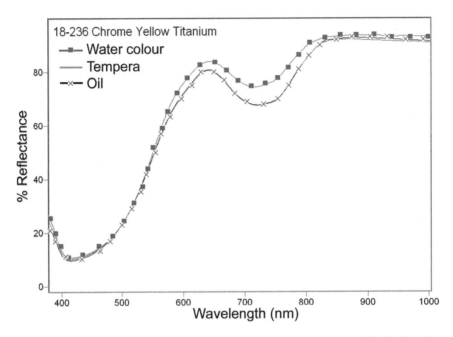

Fig. 16. The effect of binding medium on the reflectance curve – Schmincke chrome yellow titanium.

possible buckling in the canvas when laid flat (Martinez *et al.*, 2002). If mosaicing is required for larger images, the camera must be moved on an *X–Y* stage (Carcagni *et al.*, 2005), or the painting is moved step-wise on a motorised easel. Spatial resolution is determined by pixel size, distance to the object, and field of view, and refers to the size of the smallest object that can be resolved on the image.

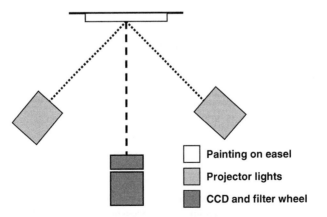

Fig. 17. Hyperspectral imaging system setup.

The image should be of sufficiently high resolution to allow all details of the painting to be analysed, including craquelure. There is no recommended pixel-per-millimetre guide-line; however, previous studies have focussed on small sections of paintings by using between 6 (Bonifazzi, 2005) and 18 pixels/mm (Scholten *et al.*, 2005), and up to 40 pixels/mm for areas with a fine network of cracks (Saunders and Cupitt, 1993). While placing the camera closer to the painting increases the pixels/mm value, the field of view is reduced, and the area of interest is subdivided into a large number of images. With pixels of size 9 µm or less, as encountered on the SBIG camera, the resolution is considered simi-lar to that of photographic colour film (Greiner, 2000). As the resolution of the image depends on the camera-to-object distance, 150 cm achieves a resolution of 4.34 pixels/mm. This was deemed sufficient for most small paintings studied, with a field of view of 276 mm × 414 mm. The resolution may be increased by moving the camera closer, *e.g.* 6.58 pixels/mm at 100 cm. The benefit of working from a longer distance is that the entire painting can be captured in one image. The importance of keeping both easel and camera perfectly aligned was demonstrated when calculating pixels per millimetre; the slightest forward or backward tilt of the grid plane caused distortion (lengthening/shortening) of the vertical axis. Both camera and surface measured must thus be level and parallel. While this may be corrected by geometric transformation and pixel interpolation, this step becomes unnecessary through correct and consistent camera placement. The camera is therefore positioned in the same vertical axis as the painting, using a sturdy tripod so that the positioning grid surrounding the area of interest fits squarely onto the monitoring screen.

3.6.3. Calibration

With appropriate calibration, the above CCD and filter system become a spatial spec-trophotometer (Baronti *et al.*, 1997; Imai *et al.*, 2000; Bell *et al.*, 2002; Day, 2003a; Roselli and Testa, 2005). To overcome variable response due to a combination of unequal inten-sity in the QTH source, unequal transmission of filters, and sensitivity of the CCD chip, basic calibration must be performed for all image collection events. For calibration, a 99% nominal reflectance Spectralon tile of 10 cm^2 is included in each image. Spectralon stan-dards have a known reflectance constant over the range 250–2500 nm with a flatness of within ± 4% (Baronti *et al.*, 1998). A background image, or dark frame (0%), is also made before each image by blocking the lens. These images are subtracted from the rest of the image to improve signal to noise. Using the spectral reflection reference tile and corrected adjustments, reflectance (R) can be computed for each pixel of the image:

$$R_{sample} = \frac{P_{reference} \times (V_{sample} - V_{dark})}{(V_{reference} - V_{dark})}$$

where P the certified spectral reflectance of a neutral diffuser, *e.g.* 99%, and V_{sample}, V_{dark}, and $V_{reference}$ are the detector signals for the captured sample, dark frame, and refer-ence, respectively. V_{dark} and $V_{reference}$ permit instrument adjustment and must be repeated before each session (Carcagni *et al.*, 2005). Calibration is thus achieved by measuring

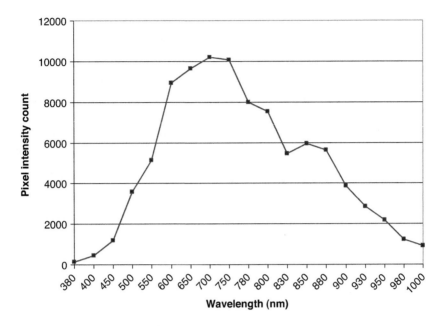

Fig. 18. Reflectance of 99% Spectralon plate using 0.12-s exposure at all wavelengths.

the dark frame and spectral reflection factor of the known Spectralon reference (Casini *et al.*, 2002b).

Based on the shortest exposure time permitted by the CCD camera, of 0.12 s, the intensity value of the Spectralon reference plate can be measured to provide the emission curve of the light source (Fig. 18). Calibration is effected by means of adjusting exposure times, which is the only easily controlled variable. The highest $V_{reference}$ is achieved between 700 and 750 nm (10 102 intensity value at Spectralon plate), so that this is assigned the shortest exposure time of 0.12 s.

Exposure times required for other wavelengths are calculated mathematically to achieve the same maximum reflectance value, and are confirmed through testing. The exposure function consists of the following algorithm:

$$\text{Exp} = \left(\frac{\text{Maximum intensity}}{\text{Intensity measured at given } \lambda} \right) \times \text{Minimum exposure time}$$

$$= \left(\frac{10102}{\text{Intensity measured at given } \lambda} \right) \times 0.12 \text{ s}$$

Here, Exp is the calculated exposure time in seconds required to equalise response maxima. From this, calibrated exposure times are calculated as listed in Table 1.

Table 1. Calibrated exposure times for the CCD camera

Filter	λ (nm)	Time (s)
1	400	2.69
2	450	1.10
3	500	0.37
4	550	0.25
5	600	0.14
6	650	0.14
7	700	0.12
8	750	0.12
9	780	0.16
10	800	0.17
11	830	0.24
12	850	0.23
13	880	0.24
14	900	0.32
15	930	0.44
16	950	0.61
17	980	1.06
18	1000	1.44

A steady CCD response is thus achieved, and by assigning the Spectralon plate as the 100% reference standard, results can be replicated across images and also directly against UV–Vis–NIR spectrophotometer results. Exposure at the shorter (400–450 nm) and longer wavelengths (980–1000 nm) require longer times than those in the mid region to overcome lower transmission of the filters and lower QE at those wavelengths.

This is because the transmittance of the filters at these wavelengths is lower (Fig. 14), the light source has lower radiance, and the CCD sensor has lower quantum efficiency at shorter wavelengths (Fig. 10).

3.6.4. Preprocessing

Digital imaging permits the correction of geometric and spectral distortions, filtering, enhancing contrast, and applying transformations to maximise information content (Prost, 2001). Some squaring of images had to be undertaken where misaligned filters caused problems when compiling the image cube. Images were grid referenced with help of a reference plate and grid, so that the target images had the same grid coordinates in all spectral bands. This reference consisted of a cardboard frame with printed regular check pattern, which was captured in every image. This provided four reference points at the image corners that could be used for skewing and cropping. Misregistration of images tended to be negligible. Integration of the results is obtained by overlapping the images.

The separate monochrome images at different wavelengths are stacked into a single high-resolution data cube (Fig. 1). Reflectance spectra for every image pixel can then be extracted by plotting the pixel reflectance values for each grid point against their frequency value (Saunders and Cupitt, 1993).

3.6.5. Spectral matching

Spectral matching is achieved in two ways. One approach to analysing a hyperspectral image is to attempt to match each image spectrum individually to one of the reference spectra in a library. This works best if there are extensive areas of essentially pure materials with corresponding reflectance spectra in the library. Alternatively, suspected colorants may be included in the image, alongside the target painting (Scholten *et al.*, 2005). This permits more rapid and direct comparison.

All of these approaches perform a mathematical transformation of the image spectra to accentuate the contribution of the target spectrum while minimising the background. The required mathematics consist of spectral matching algorithms (Wilcox, 1997; Berns and Imai, 2002). Various metrics are used for spectral matching, which attempt to find the best-fit similarity. While most early techniques were based on peak picking, search algorithms now use full-spectra searches. Because the entire region is used in the matching, this full spectrum match is also more suited to reflectance spectra that tend to have broad features rather than the sharp peaks observed in Raman spectroscopy. A number of metrics are used to match the full observed spectrum with either an identical library spectrum or the most similar library spectrum.

3.6.6. Evaluation

Spectral accuracy of results were confirmed in two ways: first, by application of Raman spectroscopy to ensure that reference standards were accurate. Some smaller paintings could also be tested for pigment composition in this manner. Second, reflectance spectra were confirmed using a Varian Cary 5 UV–Vis–NIR Spectrophotometer with diffuse reflectance accessory (DRA).

Parameters for the Cary 5 Spectrophotometer are as follows:

○ Wavelength range	300–1000 nm
○ Collection mode	Percentage reflectance (0–100)
○ Average time	1.00 s
○ Data interval	1.00 nm
○ Scan rate	60 nm/min
○ SBW	2.00
○ NIR energy level	3
○ UV/VIS energy level	1
○ Beam mode	Double
○ Slit height	Reduced

○ Source changeover UV	350 nm
○ Detector changeover NIR	870 nm
○ Grating changeover NIR	870 nm
○ Zero/baseline correction	Spectralon 99% standard
○ Software	Cary WinUV 3.00

This instrument collects spectra at 1 nm wavelength intervals, rather than every 50 nm used for the imaging system. This increased the number of data collection points is permitted by a monochromator filter, which provides a clearer reference spectrum for comparison. The same spectral region was covered (400–1000 nm) using the same Spectralon standard for baseline correction. This permitted direct comparison of spectral results. Spectra were recorded in reflectance mode, the measurements are expressed in a way analogous to absorbance: $A' = \log(1/R)$, where R is the diffuse reflectance, which allowed for easier comparison with data reported in literature. The use of a single monochromator in conjunction with a multichannel detector enhances the sensitivity of the spectrophotometer, being important given the low intensity of the diffuse reflected light.

Accuracy of spectral results may furthermore be assessed by using the following techniques:

- **Absolute reflectance error** (Imai *et al.*, 2000; Day, 2003b): This is computed as the difference between measured and predicted spectral reflectance for each wavelength. The range is then defined as the difference between the maximum and minimum of the absolute spectral error. The smaller the range of absolute error, the more the reconstructed spectrum adheres to the curvature of the measured reflectance. This method is insensitive to lightness errors, which are shown as vertical spectral shifts. A tolerance of 10% was deemed acceptable.
- **CIE L*a*b** (Imai *et al.*, 2000; Martinez *et al.*, 2002): Using and comparing CIE Lab colour space figures using CIE illuminant A and calculated for standard observer at 2°.
- **Root-mean-square** (RMS) spectral error (Melessanaki *et al.*, 2001) between each target sample's measured and estimated reflectance spectrum.

4. SELECTED APPLICATIONS

To demonstrate the power of the technique developed, the hyperspectral imaging system is applied to selected examples. First of all, the system was tested by comparing results against spectra collected by a UV–Vis–NIR spectrophotometer. When satisfactory results were achieved and replicated accurately, the reflectance spectra of many different pigment and binding media combinations were collected. During this work, it was noted that the colour blue was of particular interest to conservators, and was therefore tested for its discriminating power first on a test painting, and later on paintings on loan from the Australian War Memorial.

4.1.1. Comparing results with Cary DRA spectra

To verify the results, the reflectance spectra of the 27 Gamblin Conservation Colours were compared to those collected using a higher spectral resolution spectrophotometer. Using the Cary 5 DRA, all the Gamblin paints were measured using an integrating sphere spectrophotometer with specular component excluded, approximating the hyperspectral imaging system's lighting geometry as much as possible. Next, results from both were superimposed using Microsoft Excel spreadsheets. As can be seen in the examples provided in Figs. 19–22, the shape and percentage reflectance of reflectance spectra matched reasonably well across different colours. This demonstrates that using only 18 wavelength filters between 400 and 1000 nm, compared to the 700 wavelength points collected by the Cary instrument, is sufficient to discriminate among the paints tested.

Any deviations in spectrum are possibly attributed to the different angles of measurement. In hyperspectral imaging, this angle is 90 to the surface of the plane, whereas inside the DRA accessory of the UV–Vis–NIR spectrophotometer, excitation light hits the surface at 90°, while it is collected at 45° by the detector. The 18 wavelength collection points, however, proved to be adequate for reproducing a spectrum generated by UV–Vis–NIR spectrophotometer, and are sufficient to achieve discrimination between pigment types as further demonstrated in the following applications.

4.1.2. Collecting reference reflectance spectra

Reflectance spectra were collected for a wide range of pigments and binding media. Both Schmincke pigments mulled up into various binding media and Gamblin retouching paints

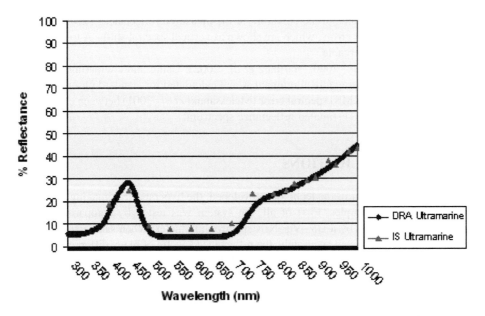

Fig. 19. Gamblin ultramarine – DRA and IS techniques compared.

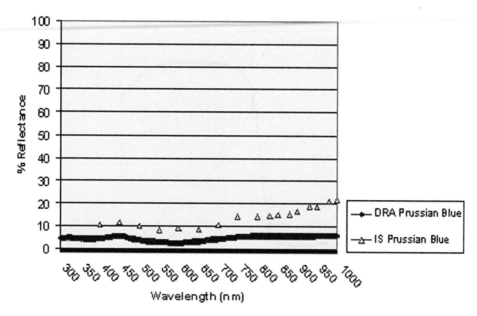

Fig. 20. Gamblin prussian blue – DRA and IS techniques compared.

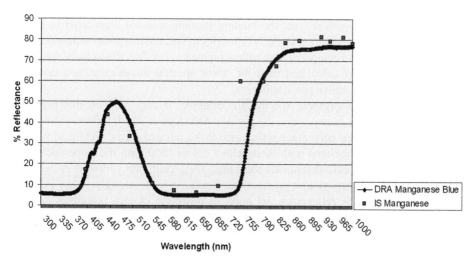

Fig. 21. Gamblin manganese blue – DRA and IS techniques compared.

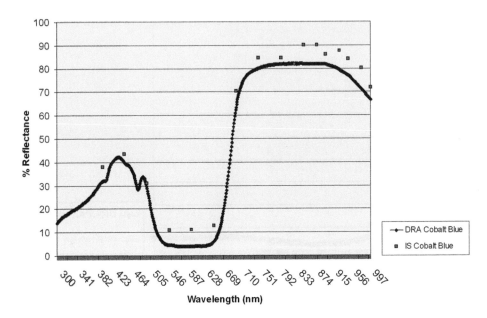

Fig. 22. Gamblin cobalt blue – DRA and IS techniques compared.

were tested in this manner. As can be seen in Fig. 23, the binding medium had little effect on shape of the spectrum; however, the pigment ground in oil was consistently found to have the lowest percentage reflectance, seen also in a comparatively darker tone of the paint. This finding is confirmed by Johnston (1973). As shape, rather than absolute value, was indicative of the pigment type, the binding medium was not considered to have a significant effect on identification work. A full series of spectral images of different colour paints were collected and compared, *e.g.* yellow pigments (Fig. 24). Reflectance spectra of the model samples were calculated from the stored spectral image cube, and saved for potential use in a comparative spectral library.

In comparing yellow pigment spectra, for example, it may be seen that pigments differentiate themselves in two ways: the position and slope at which reflectance dramatically increases, and their percentage reflectance in the infrared region. While all yellows reflect in the region associated with their yellow colour (550 nm), the slope and wavelength at which this occurs determines their hue. Thus, cadmium yellow deep is easily identified among the nine yellow pigments as it reflects most strongly in the infrared, levelling at around 90% reflectance. Its point of inflection closely resembles brilliant yellow, although the latter reflects only around 70% in the infrared. Priderite yellow has a rise in reflectance much earlier than other yellows (at 450 nm), which accounts for its green hue. Ferrite yellow and yellow ochre reveal the same reflectance pattern. This agrees with both pigments consisting of goethite, while the former is a man-made version. Other colours may be discriminated based on their characteristic reflectance spectra in the same way.

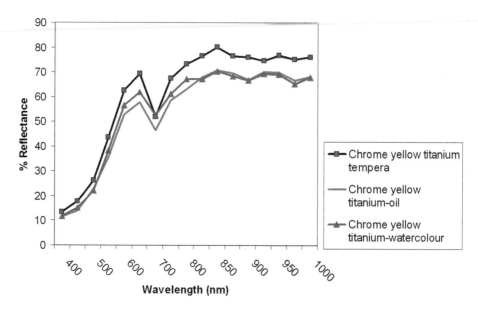

Fig. 23. The effect of binding medium on reflectance spectra – Chrome yellow titanium.

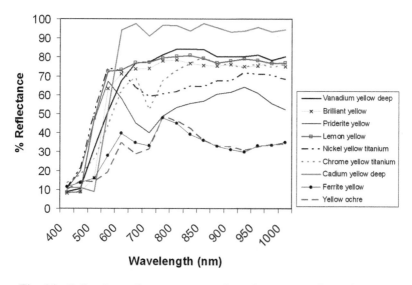

Fig. 24. Collecting reflectance spectra by colour, *e.g.* yellow pigments.

4.1.3. Application 1: blue paint

Identifying blue paints is important to the paintings conservator when choosing the right colour for retouching (Morgans, 1982; Berns and Imai, 2002; Day, 2003a; van der Weerd *et al.*, 2003; Clarke, 2004). Blue, in particular, is difficult to match, and choosing the incorrect pigment can result in severe metamerism. Pigment identification was tested using indigo, Prussian blue, light and deep cobalt blue, cobalt turquoise, cerulean blue, ultramarine, phthalocyanine blue, and phthalocyanine turquoise in different binding media.

- **Indigo blue** is dark blue and fade prone. Indigo is found as glucoside indican on leaves of *Isatis tinctoria* (woad) and *Indigofera tinctoria* from India (McLaren, 1983). It has been exported to Europe from the twelfth century (McLaren, 1983). Synthetic indigo from *o*-nitrophenyl acetic acid has been available since 1878.
- **Prussian blue** is a synthetic pigment developed in 1704 by Dippel and Diesbach. Also called iron blue, it consists of ferric ferrocyanide ($Fe_4[Fe(CN)_6]_3$) (Barnes, 1939b).
- **Cobalt blue deep and light** were isolated in 1735, but not developed as a pigment until 1802 to replace the similarly coloured smalt (Bergen, 1986). This blue contains oxides of cobalt and aluminium ($CoO \cdot nAl_2O_3$).
- **Cobalt turquoise** is a variety of cobalt blue additionally containing chromium.
- **Cerulean blue** is paler and greener than cobalt blue, and is an artificial pigment that consists of oxides of cobalt and tin ($CoO \cdot nSno_2$) (Johnston-Feller, 2001).
- **Ultramarine (French)** is an artificial substitute for the expensive natural ultramarine derived from lapis lazuli (Tate, Paint and Painting, 1982). It was discovered in 1826, but not commercialised until 1830 by Guimet and Koetting (Bergen, 1986).
- **Phthalocyanine blue** was prepared by Von Diesbach and Von der Weid in 1927 (Loebbert, 1992). Its exceptional stability, lightfastness, and resistance to acids, alkalis, and heat led to copper phthalocyanine being commercialised as a pigment by ICI in 1935, with IG Farben and DuPont starting soon after. Winsor and Newton produces phthalocyanine blue under the name Winsor blue (Tate, Paint and painting, 1982).
- **Phthalocyanine turquoise** is a mixture of phthalocyanine blue and phthalocyanine green ($C_{32}H_2Cl_{15}CuN_8$).

Reference spectra were collected directly from these common blue paints seen in Fig. 25. As seen with the yellow pigments, the nine blue pigments separate easily, based on both different value and shape reflectance peaks in the blue region (400–450 nm), and different levels of reflection in the infrared (Fig. 25). Most pigments reflect light in the blue region of the spectrum and absorb well into the red region. With the exception of phthalocyanine blue, phthalocyanine turquoise, and Prussian blue, all blues show an upwards slope at 780 nm. Cobalt blue deep and cobalt blue light may only be separated by percentage reflectance peaks at 450 nm. Some of the deep blue/black pigments, Indigo in particular, extinguishes the incoming beam in the visual region to a large extent, but the intensity of the long-wavelength reflection of the spectrum (700–750 nm) is high. The deep colour of Prussian blue is explained by its low reflectance curve, which reflects little light between 400–1000 nm. This reflection spectrum is almost featureless. Some of the spectra were quite close in shape and were therefore difficult to separate by spectrum (*e.g.* cobalt blue light and cobalt blue deep). These results are confirmed by van der Weerd *et al.* (2003) and Berns and

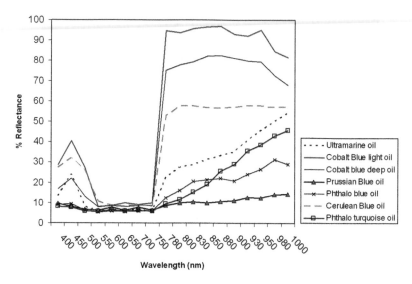

Fig. 25. Reflectance spectra of representative blue paints.

Imai (2002), who also found blues easy to separate. Berns and Imai incorrectly identified manganese blue as phthalocyanine blue, which may have been due to similar incorrect labelling of pigments.

Blue pigments also provided the opportunity to measure CIE $L*a*b*$ colours using Thermo Grams AI©. This program converted reflectance spectra using 2° observer angle and standard illuminant CIE A, and calculated CIE $L*a*b*$ values between 400 and 830 nm (Table 2). This demonstrates how easily CIE $L*a*b*$ values are determined, useful for recording colour change in an objective manner.

Table 2. CIE $L*a*b*$ values derived from reflectance spectra of blue paints

Paint	$L*$	$a*$	$b*$
Ultramarine blue oil	24.005	5.356	−43.318
Ultramarine blue tempera	29.673	3.576	−72.191
Ultramarine gamblin	25.130	7.061	−47.966
Cobalt blue deep tempera	29.530	−0.123	−54.090
Cobalt blue gamblin	31.121	1.450	−60.498
Cobalt blue light tempera	45.237	−6.571	−61.429
Prussian blue tempera	12.402	3.004	−14.658
Prussian blue gamblin	19.770	0.072	−11.424
Indigo oil	25.690	2.008	0.395
Manganese blue gamblin	40.015	−27.707	−54.071
Phthalo blue oil	23.443	−1.582	−12.553

Fig. 26. Blue painting with Schmincke reference plates.

4.1.4. Application 2: gouache painting in blue

To test the identification capacity for blue paints, the system was applied to a watercolour and gouache painting made by the author (Fig. 26). This also validated methodology before field testing. A number of blue Artist Spectrum (Australia) gouache paints and Windsor and Newton (UK) watercolours were combined into a single portrait of a lady. Schmincke reference plates were used to attempt matching and identification of the "unknowns". Figure 27 presents the location of these gouaches and watercolours that were tested.

All blues present in the reference set were identified (Figs. 28–32). Where overlap was present between reference "known" paints and the "unknowns" of the painting, successful matching was achieved, *e.g.* phthalocyanine blue (Fig. 31) and ultramarine (Fig. 28). Locations for these identified paints mapped well across the image. While this is normally seen as red highlights against a black and white image, reproductions in this chapter do not permit colour and instead locations are mapped as bright white areas. Cerulean blue (Fig. 29) had

Ultramarine

Cobalt Blue

Prussian Blue

Cerulean Blue

Zinc White

Turquoise

Marine Blue

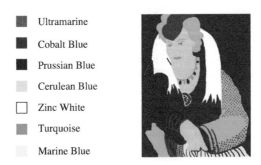

Fig. 27. Location of Gouache and watercolour "unknowns".

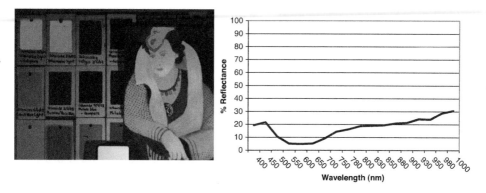

Fig. 28. Ultramarine locations and spectrum.

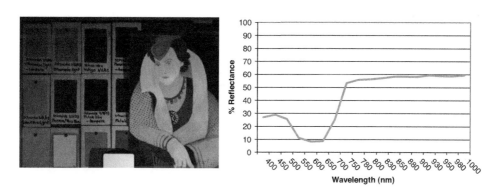

Fig. 29. Cerulean blue locations and spectrum.

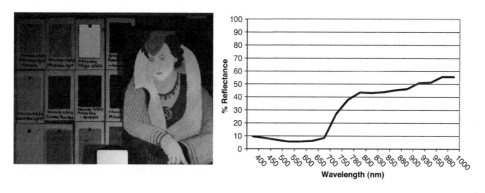

Fig. 30. Indigo blue locations and spectrum.

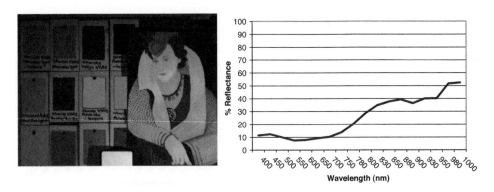

Fig. 31. Phthalocyanine blue locations and spectrum.

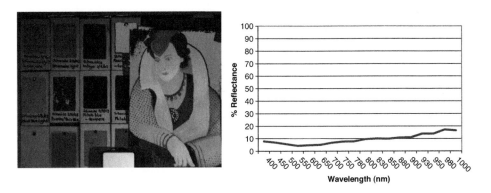

Fig. 32. Prussian blue locations and spectrum.

been used in a diluted manner across the face and hands; however, the diluted spectrum of these areas was not strong enough for identification. Limits were further presented where reference spectra were not available for particular paints present, *e.g.* turquoise blue (Fig. 33), marine blue, and cobalt blue. This highlights the need for either an extensive reference set to be included in the analysed image, or a searchable library database of reflectance spectra to be accessible for comparison.

4.1.5. Application 3: small war memorial paintings

With the assistance of David Keany, senior paintings conservator of the Australian War Memorial, two studies in oil measuring 7 cm × 12 cm were provided on loan for further investigation (Fig. 34). These were deemed suitable due to their small size, and the apparently large variety of blues employed.

The locations of pure Prussian blue could easily be identified by comparison of a reference plate in the same image (Figs. 35 and 36). Mixtures using lead white to obtain lighter

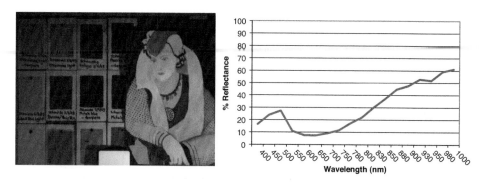

Fig. 33. Turquoise blue locations and spectrum.

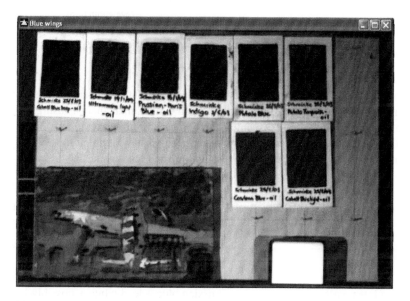

Fig. 34. Small war memorial painting (*Wings*) with selected blue reference plates.

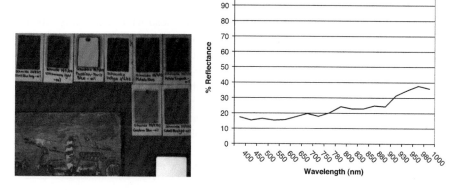

Fig. 35. *Wings* Prussian blue locations and reflectance spectrum.

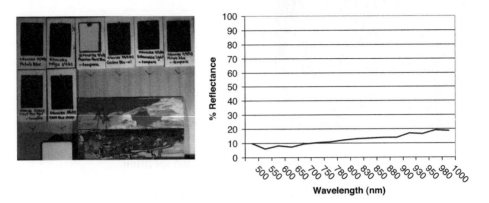

Fig. 36. *Plane* Prussian blue locations and reflectance spectrum.

shades of blue are further investigated. The blue pigment composition was confirmed using Raman spectroscopy (Lee *et al.*, 2005). Results of Raman analysis using the Kremer database are presented in Table 3, while sample locations are given in Fig. 37. It appears the artist has used a very simple palette: the lead white is almost ubiquitous, while all blue areas consisted of various mixtures of Prussian blue and lead white.

Table 3. Location and identification of AWM plane pigments using Raman spectroscopy

Sample	Description	Identification
P1	Blue sea	Prussian blue
P2	White cloud	Lead white, small amount of Prussian blue
P3	Green earth	Prussian blue + lead white + possibly kibeni orange or priderite yellow
P4	Light brown ground	Lead white, other inclusions seen but no spectra generated.
P5	Medium blue	Prussian blue
P6	Medium blue sky	Prussian blue, lead white
P7	Green foreground	Prussian blue + lead white + possibly kibeni orange or priderite yellow
P8	Purple pants	Prussian blue
P9	Pink man on wing	Lead white
P10	Red hair on LH man	Possibly kibeni orange or priderite yellow
P11	Red hair on LH man	Possibly kibeni orange or priderite yellow
P12	Dark blue wing	Prussian blue, small carbon peaks indicating carbon black
P13	Light brown fabric coating	Lead white
P14	Light blue	Lead white, low amount of Prussian blue

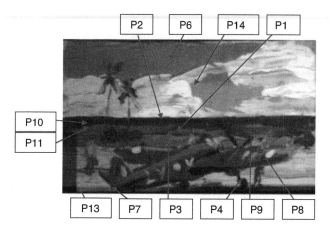

Fig. 37. AWM plane locations for Raman sampling.

These dark blue regions have also been identified as Prussian blue using imaging spectroscopy. This shows that identification by imaging spectroscopy has been partially successful in identifying the blue paint used, although lighter mixtures using lead white have made this less than straightforward.

4.1.6. Application 4: Sidney Nolan

In order to test the portability and effectiveness of the instrument setup, the imaging system was taken to the Australian War Memorial (Fig. 45). Here, it could be tried with ambient lighting conditions and actual paintings with unknown pigment composition. It also allowed for field testing, seeing how portable the instrumentation was and how easy it was to carry and set up in a museum location with uncontrolled lighting. This was an important consideration for the system's design. Application proved to be very straightforward, with the entire system being able to be dismantled, carried by hand, and fitted in the back of a car. The War Memorial venue consisted of an active conservation studio, with diffuse natural light, fluorescent light, and incandescent lights, simultaneously. As in the lab, the camera was positioned at 150 cm from the painting, which was sufficient to capture the entire image. Projectors could be placed at slightly longer distances as the allowed space was wider, in this case 120 cm compared to 90 cm. This affected maximum exposure readings for the Spectralon reference, so that calibration had to be performed before images were captured.

The Angry Penguins group was of particular interest, with little materials research performed to date on some of Australia's best known artists, *e.g.* Sydney Nolan (1917–1992). This has led to the formulation of incorrect hypotheses about his materials, and incorrect labelling in catalogues (Kubik, 2007). Nolan was unusually experimental in his approach to paints, in part due to influence from overseas artists, but also caused by rationing and cost of materials during the early stage of his career. Returning to interviews (Nolan, 1962; 1980; 1987) and letters written by Nolan (Nolan, 1943b; Nolan, 1943c; Nolan, 1943d; Nolan, 1943e; Nolan, 1947; Nolan, 1948) reveals what may be expected in materials analysis.

In his letters, there are several requests by Nolan for a particular blue dye (Nolan, 1943f), which has never been identified. By 1954, Nolan had switched to mixing his own pigments into polyvinyl acetate binder (Llyn, 1967). Advantage was taken of the War Memorial's collection of Nolan paintings to shed light on these details. The following three paintings containing blue paint were tested:

(1) *Gallipoli Landscape, half-lit* (1958): Synthetic polymer paint on coated card, ART 91243
(2) *Cove at Hydra* (1956): Synthetic polymer paint on coated card, ART 91261
(3) *Gallipoli Landscape with gun* (ca. 1963): Acrylic with yellow oil crayon on gloss art board, ART 90217

All three contained unknown blue paint in either sky or sea, so that these could be tested against known references and the library database created earlier. The blues also appeared in different hues, so that several types of pigments were expected.

The first painting tested was *Gallipoli Landscape, half lit*. This showed dappled blue paint, almost ultramarine in colour. Nolan's technique at this stage involved wiping paints using a squeegee and sponge (Moore, 1998). The painting was placed on an easel, and projector lights placed at 1.0 m at 45° incident to the surface plane of the painting. As had been established at ANU, the CCD camera was placed at 150 cm from the painting, which was sufficient to capture the entire image, as well as six blue reference plates and the Spectralon reflectance standard in the middle of these (Fig. 38). The positions of the blue references remain consistent throughout the three Nolan paintings tested. The white balance was calculated at 4000 maximum and the exposure times adjusted accordingly. Images were captured sequentially between 400 and 1000 nm, and stacked into a data cube

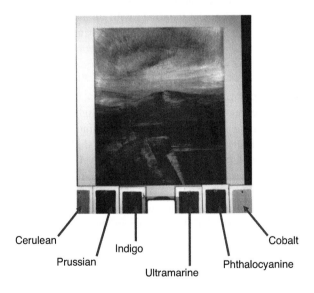

Fig. 38. Nolan's *Gallipoli Landscape, half lit* with six reference plates of blue pigments and the Spectralon standard.

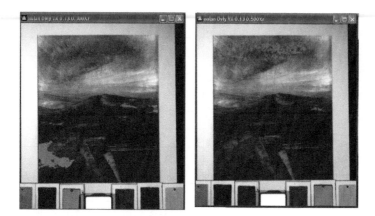

Fig. 39. *Gallipoli Landscape, half lit* locations for indigo (left) and phthalocyanine blue (right).

using the Hypercube software. From this, identification could either be achieved by selecting the six standards and finding similarities within the image, or reversely by selecting key points in the sky and allowing the search to find similarities within the reference plates.

After collating the images in Hyperspec and searching both the distribution of unknown blue areas and the references, both indigo and phthalocyanine blue were found to be present as a mixture (Figs. 39 and 40). The two different colours appeared thoroughly mixed on the canvas, and were only discernable using imaging spectroscopy. This was therefore a positive outcome, as it indicates the possibility of identifying individual components within a mixture.

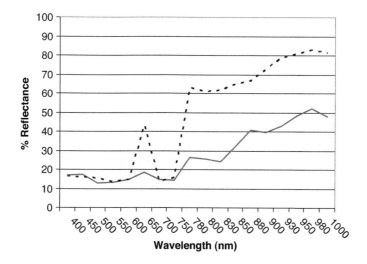

Fig. 40. Reflectance spectra of indigo (dotted line) and phthalocyanine blue (solid line).

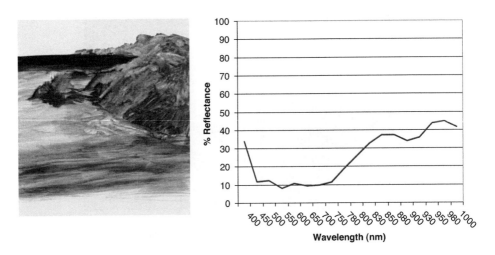

Fig. 41. Sidney Nolan (1956) *Cove at Hydra.*

The second Nolan to be tested was *Cove at Hydra*, with much darker blue paint present in the sea (Fig. 41). This painting was interesting as the paint was solidly applied at the horizon line, and gradually washed out towards the bottom using a diluent, becoming paler in tone as the white ground was allowed to show through. This permitted testing on different concentrations of the same paint, which is similar in effect to mixtures with white. As with *Gallipoli Landscape*, the dark blue was found to be phthalocyanine blue (Fig. 42).

Fig. 42. Phthalocyanine blue reflectance spectrum and distribution.

Fig. 43. Sidney Nolan (c1963) *Gallipoli Landscape with gun.*

Distribution for this reference spectrum, however, was limited to the dark blue in the background, and not the diluted areas in the foreground. A similar effect was observed with diluted cerulean Blue.

The third Nolan painting to be tested was *Gallipoli Landscape with Gun* (Fig. 43). The blues in this third painting appear to consist solely of indigo blue, seen distributed in the foreground and left sky (Fig. 44).

Thus, it may be concluded that Nolan used a combination of indigo and phthalocyanine blues in the three paintings studied. Indigo and phthalocyanine could easily be identified

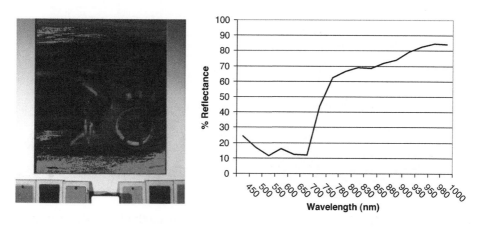

Fig. 44. Indigo blue, distribution and reflectance spectrum.

Fig. 45. Imaging system set up at the Australian War Memorial's Treloar offices, with data capture of Ivor Hele.

separately, based on their reflectance spectra, which also matched those collected during blue reference studies. As before, the spectrum of indigo shows a much higher reflectance in the NIR range, whereas phthalocyanine is practically absorbent throughout the 400–1000 nm range.

4.1.7. Application 5: Ivor Hele and a copy

Also, at the Australian War Memorial, it was possible to compare a recent copy against the original painting by Ivor Hele–*Private John Growns* (1952) (Fig. 46). As the artist of the copy, John Ballan, was still in contact with the AWM, it was possible to ask which

Fig. 46. Ivor Hele (1952) *Private John Growns*, and Ballan (2005) *Copy of Private John Growns*.

pigments he had used for his copy. According to Ballan, his palette consisted of the following pigments:

–Lead white
–Yellow ochre
–Mars red
–Alizarin crimson
–Raw umber
–Payne's grey
–Cadmium yellow
–Cadmium red

The captured images of the Ballan copy produced a good-quality data cube, with a flat response from the Spectralon cube, and therefore a good-quality spectra for analysis and comparison. For identification, a Gamblin target containing most of the above pigments was captured under the same lighting conditions to replicate pigment spectra more accurately. This permitted searching for distribution of known pigment spectra and possible identification of paints used within the two target images.

Spectra were extracted from different areas of both paintings and compared (Figs. 47–50). From this, it may be seen that similarly coloured regions revealed dissimilar spectra and distributions, easily highlighting the use of different paints and techniques, *e.g.* facial colour, seen in Figs. 47 and 48, where Ballan has extended the use of this paint into the forehead and right shoulder. The two spectra could not be identified against standards, but are possibly a combination of yellow ochre and cadmium red, with a yellow peak at 600 nm evident in some of the more yellow areas (Fig. 51), and a flat, high reflectance between 750 and 1000 nm, characteristic of cadmium paints (Fig. 24). In the copied version by Ballan, raw umber was identified in the dark brown regions of the hat, with a small reflectance maximum at 700 nm (Fig. 49). The distribution of this pigment on its own appears limited. When testing similarly coloured dark areas of the Hele painting (Fig. 50), distribution of this colour is more widespread. The spectrum of this region has a

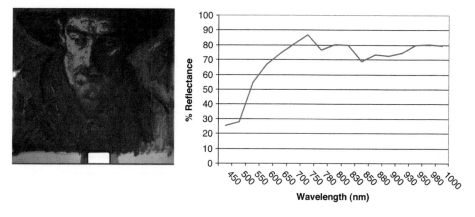

Fig. 47. Ballan results: facial colour distribution and spectrum.

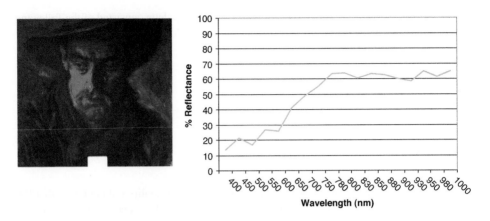

Fig. 48. Hele results: facial colour distribution and spectrum.

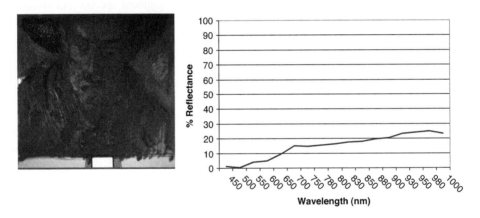

Fig. 49. Ballan results: limited umber distribution and spectrum.

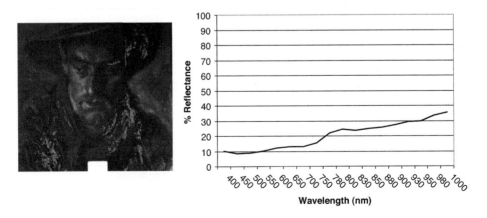

Fig. 50. Hele results: shadows distribution and spectrum.

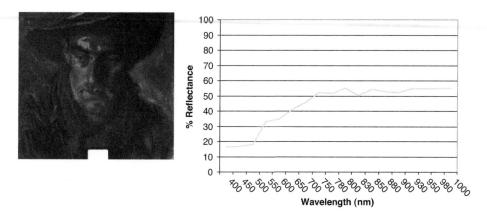

Fig. 51. Hele results: highlights distribution and spectrum.

reflectance peak at 800 nm, so that this colour must be attributed to a different pigment or mixture.

Other colours tested, such as the green uniform and purple background, could not be identified. In fact, few pigments could be spectrally matched due to the complexity of spectral mixtures. Instead, spectra revealed working methods and allowed comparison between the two paintings to see whether the same mixtures had been used. Dark brown areas varied between the two artists, and similar areas or spectra could not be found. The raw umber regions also did not match those of the shadows, meaning a different dark pigment had been used. While similarly shaped spectra were produced by both portrait's face colours (Figs. 47 and 48), a slight variation in reflectance peak maxima was observed, possibly due to the contributing effects of a mixture.

The technique is therefore useful to discriminate similarly coloured areas by distribution and reflectance spectra, but the complexity of mixture analysis is evident. An unexpected result was found by comparing reflectance images at 1000 nm. Here, the two paintings were easily differentiated as Ivor Hele had resorted to extensive underdrawings before commencing painting (Fig. 60), while Balan had not make preparatory sketches.

4.1.8. Application 6: documenting and monitoring varnish

UV light is not useful for pigment identification as it does not penetrate below any surface coating such as dirt, dust, or varnish. Instead, it offers the opportunity to detect and monitor the yellowing of such varnish layers present. Detecting varnish on a painting has typically been achieved using UV light fluorescence and documented using a normal camera with UV cutoff filter (such as the Kodak 2E). This may similarly be achieved using the hyperspectral imaging system with only three interference filters. A colour composite is created using the three wavelengths 450, 550, and 650 nm to create a true colour replicate of any fluorescence detected (Fig. 52). The various features typically detected in this manner are thus easily revealed, including breaks in the varnish layer caused by a tear

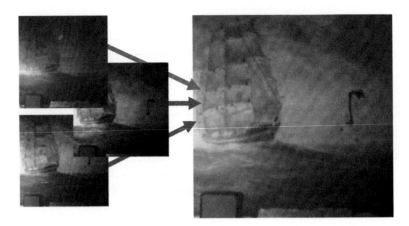

Fig. 52. A colour composite image of varnish fluorescence is created by combining the three images at green, blue, and red wavelengths.

repair, retouchings, a test varnish removal section, and several areas of losses. To achieve correct white balance, equal reflectance must be achieved for the three colours using the Spectralon plate. This is achieved by varying the exposure times.

Also, the yellowing of varnish may be monitored by collecting reflectance spectra. To test this, a partially cleaned painting with aged dammar varnish (Fig. 53) was tested. The central yellowed stripe represents the remaining varnish, while the areas to the left and right of this have been cleaned using solvents. Reflectance spectra collected from the two areas could then be compared (Fig. 54).

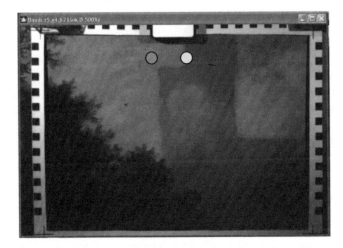

Fig. 53. Partially removed varnish with tested areas (blue dot = cleaned area, yellow dot = aged varnish).

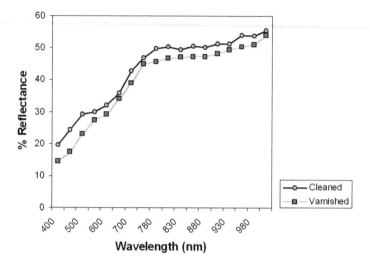

Fig. 54. Reflectance spectra of cleaned and varnished areas.

Comparing the two reflectance curves, it can be seen that the varnish causes not only a drop in reflectance, but also a shift in the peak at 500–550 nm, describing the yellowed effect. A peak shift is also noted from 800 down to 780 nm.

Thus, the gradual yellowing of a varnish layer may be monitored objectively using hyperspectral imaging. This application demonstrates the advantage of this method that, in taking measurements under exactly the same experimental conditions, comparison against reference standards is straightforward, and relative measurements and differences become much more significant than with other methods.

4.2. Discussion

Based on the application and results of the new hyperspectral system, it was found to be a successful technique that was easy to apply. It is shown that a wide range of colours and pigments may be identified in this manner.

4.2.1. Ambient light

The effect of ambient light on measurements was negligible; to check, measurements were taken with projector lights only, with projector lights in combination with fluoro, and/or tungsten. This did not cause a significant change in intensity of the Spectralon plate throughout the region 400–1000 nm (Fig. 55). The system may thus be considered to be immune to the effects of ambient lighting, provided calibration using the Spectralon reference plate has been performed.

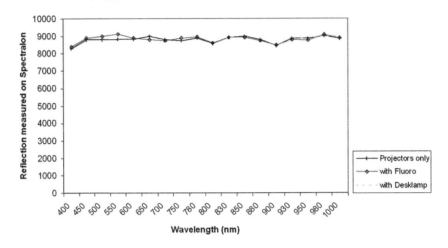

Fig. 55. The effect of ambient light on the Spectralon plate.

4.2.2. Reducing the number of filters

While in an optimal system even more bands than were used are desirable in order to cover all possible colours, an increase in filters also increases the associated cost and processing time (Day, 2003a). It also dramatically increases the difficulties in ease of use in terms of handling and flat-fielding (Slawson *et al.*, 1999). For basic analysis work, it may be possible to reduce the number of filters required. As seen in the comparison between UV/Vis/NIR spectrometer results and the IS spectra, only a few points are required to characterise the smooth spectral curves presented by artists' pigments. This is confirmed by other researchers, who were able to reduce the number of filters to 12 (Saunders and Cupitt, 1993) or even 7 (Day, 2003a).

By plotting a number of different blue pigments in a variety of media, it may be seen that discrimination between blues is already possible based on only two wavelengths, such as 780 and 1000 nm (Fig. 56). These two points represent the start and end points of increased spectral reflectance in the near infrared, which were found to be the most characteristic points in the blues tested. While this is an oversimplification, it illustrates the possible reduction in cost, capturing, and processing time. While two filters present a viable system for discriminating between most blues, other colours would require filters at wavelengths specific to them. For completely unknown pigments, many filters are required, but the system may be tailored to certain recognition tasks, selecting those bandwidths where there is most difference.

4.2.3. Mixture analysis

When the size of the measured surface is large, it is likely that more than one material contributes to the overall spectrum measured, and the mixed surface is overlapped onto a single pixel. The result is a composite or mixed spectrum, where the energy reflected from the materials combines additively (Adams *et al.*, 1993; Ichoku and Karnieli, 1996;

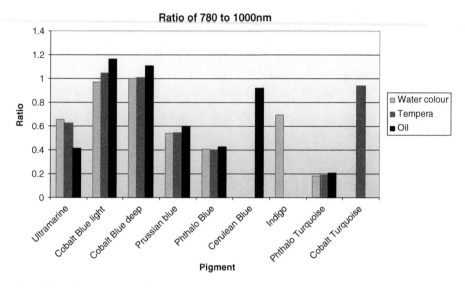

Fig. 56. Discriminating blues using only two datapoints (ratio of 780 to 1000 nm).

Roberts *et al.*, 1998; Berns *et al.*, 2005). Spectral mixture analysis (SMA) is used in remote sensing applications to calculate land cover fractions within a pixel, and involves modelling a mixed spectrum as a combination of pure land cover spectra (Bell *et al.*, 2002; Wu and Murray, 2003). The software SMA models a hyperspectral image as a linear combination of end-member spectra, and anomalous materials that do not fit the model are detected as model residuals. While opportunities for mixture analysis are offered by the Kubelka–Munk equation, a simple demonstration is presented by the simple mixture of Prussian blue and lead white seen in the Australian War Memorial's small painting of plane wings (Fig. 57).

The contribution of lead white to the lighter blues changes the spectrum in such a way that identification is not possible using only the Prussian blue reference target.

The contribution from lead white causes not only an increase of percentage reflectance in the lighter blue mixtures' spectrum, but also changes the shape of the curve (Fig. 58). This is particularly visible in a reflectance peak at 450 nm, and an upward slope in reflectance at 1000 nm. The two lighter blue mixtures thus lie in between the two pure spectra of Prussian blue and lead white, and would need deconvolution using the Kubelka–Munk equation or other mixture analysis.

4.2.4. Opportunities and interference from reflected UV and NIR

Ultraviolet light reflectance studies present an extension to the objective method to monitor colour changes such as fading. However, the presence of an aged, yellowed varnish may mask the spectra of underlying paints and interfere with their identification. As can be seen in Fig. 59, excitation light at 380 nm was completely absorbed by the surface varnish layer, preventing any studies of the layers below. Therefore, while UV light reflectance provided an extension of the reflectance curves used for pigment identification,

Fig. 57. The War Memorial's *Wings* painting consists of a mixture of Prussian blue and lead white.

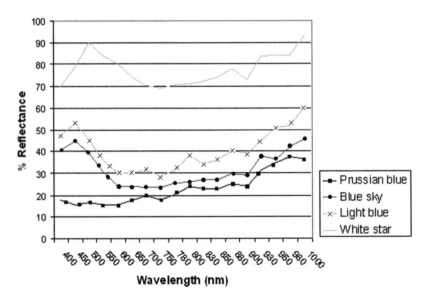

Fig. 58. Reflectance spectra of lead white (white star), Prussian blue, and mixtures of the two which generate the sky and light blues seen in the *Wings* painting.

Fig. 59. UV fluorescence colour composite and complete absorbance at 380 nm, showing how the varnish layer interferes with pigment identification.

the necessary change of source to a UVB mercury tube, the extended exposure times, and cost of extra filters in this region does not support the need for these reflectance measurements.

Conversely, while UV light was not penetrative enough, IR light near 1000 nm was able to penetrate several layers to reveal the hidden underdrawings (Fig. 60). This again is not sufficiently selective to study one paint layer at a time, and care must be taken that the reflectance measurement is specific to the layer being investigated. As underdrawings were only revealed in this painting beyond 950 nm, it is proposed that reflectance measurements be limited to 400–900 nm to focus only on the visible paint layer.

Fig. 60. NIR reflectance at 1000 nm showing underdrawings.

4.2.5. Limitations

While the system proved sound for a number of applications, there are some limitations of which the user must be aware.

- The colour of a painting can be altered with time due to, for example, the deposition of dirt particles, chemical alteration due to pollutants, photo-bleaching, and photo-oxidation. Ageing of the binding medium may also lead to sample spectra that deviate considerably from the reference. Thus, library searches may not adequately match regions with the target spectra (Mansfield *et al.*, 2002; Clarke, 2004), and further studies into the ageing of paint is recommended.
- While reflectance in the NIR region permits the study of underdrawings, penetration of the paint layer does not aid pigment identification. Instead, it may cause spectral interference from underlying materials. This did not appear to be a problem for highly reflective blues, but may be of concern for paints that are transparent in this spectral region.
- Optical glare (specular reflectance) is defined as non-image-forming light incident on the detector plane (Day, 2003b), and may be caused by extraneous reflections from the painting if this is varnished or glazed, or if lights are placed at an unsuitable angle. Glare is reduced by the use of a baffle (Imai *et al.*, 2002), covering highly reflective objects, and adjusting the position of the light sources.
- Filter angle is crucial, as the light's angle of incidence affects transmittance percentage and wavelength. Light that is off axis can have a bandwidth that is shifted as much as ¼ of the ideal value towards the red region of the spectrum (Day, 2003b). Filter angle variance was overcome by designing grub locks on all carrousel windows, and ensuring that the filter wheel was correctly mounted.
- An incomplete spectral library or reference target affects accuracy (Berns and Imai, 2002). Diagnostic information is improved by developing reference samples that emulate all the possible combinations of materials used in a painting, in order to directly compare their diffuse reflectance with those acquired from the original object. For this purpose, we have studied three paintings by Sidney Nolan for identification purposes.
- Imaging technology applied to conservation is now advancing quickly, with a number of researchers exploring the field overseas, and continuing improvements in equipment: Berns *et al.* increased the sampling image size by the use of a 5440 × 4080 pixel CCD (Berns *et al.*, 2005). Imaging spectroscopy is also being applied through microscopy (Payne *et al.*, 2005), where it is known as chemical imaging. At the same time, the number of filters is being reduced to reduce cost, labour, and the required expertise (Imai *et al.*, 2003; Carcagni *et al.*, 2005).

4.3. Conclusion

Hyperspectral imaging is an exciting new field for conservation science, with many opportunities for application. A simple, portable instrument was developed and applied to a range of applications. It was demonstrated to be successful for pigment identification, colour and UV fluorescence documentation, and basic NIR reflectance work.

While it does not offer the same level of specificity as Raman spectroscopy, hyperspectral imaging overcomes many limitations presented by the former, including that of cost, portability, and freedom from the effects of ambient light. It is a rapid technique for establishing the identification and distribution of pigments where it is desirable to localise specific pigments of known chemical composition on the work; reference spectra of appropriate pigments can be acquired, and the spectra from the painting can then be matched to the target reference spectra that may be further confirmed by the use of allied techniques. Ideally, in these situations, a spectral library search can be used.

ACKNOWLEDGEMENTS

The author is grateful to Professor Elmars Krausz (Research School of Chemistry, Australian National University) for his guidance and help during her doctoral studies at the Australian National University, where she held a postgraduate scholarship. She is grateful to the advice of Professor Dudley Creagh (University of Canberra) and to the Australian Research Council, which provided some financial assistance during the course of her studies.

REFERENCES

Adams, J., Smith, M., Gillespie, A., 1993. Imaging spectroscopy: Interpretation based on spectral mixture analysis. In: *Remote Geochemical Analysis: Elemental and Mineralogical Composition.* Cambridge University Press: Cambridge.

Ammar, H., Fery-Forgues, S., Gharbi, R., 2003. UV/Vis absorption and fluorescence spectroscopic study of novel symmetrical biscoumarin dyes. *Dyes and Pigments.* **57**, 259–265.

Attas, M., Cloutis, E., Collins, C., Goltz, D., Majzels, C., Mansfield, J., Mantsch, H., 2003. Near-infrared spectroscopic imaging in art conservation: Investigation of drawing constituents. *Journal of Cultural Heritage.* **4**, 127–136.

Bacci, M., 1995. Fibre optics applications to works of art. *Sensors and Actuators* B **29**, 190–196.

Bacci, M., 2000. UV-Vis-NIR, FT-IR, and FORS spectroscopies. In *Modern Analytical Methods in Art and Archaeology*, Wiley Interscience: New York, 321–362.

Bacci, M., Baldini, F., Carla, R., Linari, R., 1991. A colour analysis of the Brancacci Chapel frescoes. *Applied Spectroscopy.* **45**(1) 26–31.

Bacci, M., Baronti, S., Casini, A., Lotti, F., Picollo, M., 1992. Non-destructive spectroscopic investigations on paintings using optical fibres. In: Materials issues in art and archaeology III: Symposium held 27 April–1 May 1992, San Francisco, 265–283.

Bacci, M., Fabbri, M., Picollo, M., Porcinai, S., 2001. Non-invasive fibre optic Fourier transform-infrared reflectance spectroscopy on painted layers - Identification of materials by means of principal component analysis and Mahalanobis distance. *Analytica Chimica Acta.* **446**(1–2), 15–21.

Bacci, M., Picollo, M., Radicati, B., Casini, A., Lotti, F., Stefani, L., 1998. Non-destructive investigation of wall painting pigments by means of fibre-optic reflectance spectroscopy. *Science and Technology for Cultural Heritage.* **7**(1), 73–81.

Balas, C., Papadakis, V., Papadakis, N., Papadakis, A., Vazgiouraki, E., Themelis, G., 2003. A novel hyper-spectral imaging apparatus for the non-destructive analysis of objects of artistic and historic value. *Journal of Cultural Heritage.* **4**(1), 330–337.

Barnes, N., 1939a. A spectrophotometric study of artists' pigments. *Technical Studies in the Field of Fine Arts.* **7**(3) 120–138.

Barnes, N., 1939b: Color characteristics of artists' pigments. *Journal of the Optical Society of America*. **29**, 208–214.

Baronti, S., Casini, A., Lotti. F., Porcinai, S., 1997. Principal component analysis of visible and near-infrared multispectral images of works of art. *Chemometrics and Intelligent Laboratory Systems*. **39**(1), 103–114.

Baronti, S., Casini, A., Lotti, F., Porcinai, S., 1998. Multispectral imaging system for the mapping of pigments in works of art by the use of principal-component analysis. *Applied Optics*. **37**(8), 1299–1309.

Bell, J., Farrand, W., Johnson, J., Morris, R., 2002. Low-abundance materials at the Mars Pathfinder landing site: An investigation using spectral mixture analysis and related techniques. *Icarus*. **158**, 56–71.

Bergen, S., 1986. Painting technique: priming, coloured paint film and varnish. In: *Scientific Examination of Easel Paintings*. Oleffe: Court St Etienne, Belgium.

Berns, R., Imai, F., 2002. The use of multi-channel spectrum imaging for pigment identification. In: ICOM Committee for Conservation 13th Triennial meeting, Rio de Janeiro, Brazil. 217–222.

Berns, R., Krueger, J., Swicklik, M., 2002. Multiple pigment selection for inpainting using visible reflectance spectrophotometry. *Studies in Conservation* **47**, 46–61.

Berns, R., Taplin, L., Imai, F., Day, E., Day, D., 2003. Spectral imaging of Matisse's Pot of Geraniums: A Case study. In: IS&T/SID Eleventh Color Imaging Conference, Scottsdale, Arizona, 149–153.

Berns, R., Taplin, L., Nezamabadi, M., Mohammadi, M., Zhao, Y., 2005. Spectral imaging using a commercial colour-filter array digital camera. In: 14th Triennial Meeting ICOM-CC, The Hague, 743–750.

Best, S., Clark, R., Daniels, M., Porter, C., Withnall, R., 1995. Identification by Raman Microscopy and Visible Reflectance Spectroscopy of pigments on an Icelandic manuscript. *Studies in Conservation*. **40**(1) 31–40.

Bonifazzi C., 2005. Multispectral examination of paintings: A Principal component image analysis approach, in: Art '05 - 8th International Conference on Non-Destructive Investigations and Microanalysis for the Diagnostics and Conservation of the Cultural and Environmental Heritage., Lecce, Italy, A210.

Burns, R., 1993. Origin of electronic spectra of mineral in the visible to near-infrared region. In: *Remote Geochemical Analysis: Elemental and Mineralogical Composition*. Cambridge University Press: Cambridge.

Carcagni, P., Patria, A., Fontana, R., Greco, M., Mastrianni, M., Materazzi, M., Pampaloni, E., Pezzati, L., Piccolo, R., 2005. Spectral and colorimetric characterisation of painted surfaces: A Scanning device for the imaging analysis of paintings. In: Art '05 - 8th International Conference on Non-Destructive Investigations and Microanalysis for the Diagnostics and Conservation of the Cultural and Environmental Heritage. Lecce, Italy, A117.

Casini, A., Lotti, F., Picollo, M., Stefani, L., Aldrovandi, A., 2002a. Fourier transform interferometric imaging spectrometry: A new tool for the study of reflectance and fluorescence of polychrome surfaces. In: Conservation Science 2002. 22–24 May, Edinburgh, Scotland, 249–253.

Casini, A., Radicati, B., Stefani, L., Bellucci, R., 2002b. A multispectral scanning device for reflectance and colour measurements: An Application to the Croce di Rosano. In: Art 2002: 7th International Conference on Non-destructive Testing and Microanalysis for the Diagnostics and Conservation of the Cultural and Environmental heritage. Antwerp, Belgium.

Clarke, M., 2004. Anglo-Saxon manuscript pigments. *Studies in Conservation*. **49**, 231–244.

Cordy, A., Yeh, K., 1984. Blue dye identification on cellulosic fibers: Indigo, logwood and Prussian blue, *Journal of the American Institute for Conservation* **24**(1), 33–39.

Curiel, F., Vargas, W., Barrera, R., 2002. Visible spectral dependence of the scattering and absorption coefficients of pigmented coatings from inversion of diffuse reflectance spectra, *Applied Optics*. 41 **28**, 5969–5978.

Day, E., 2003a. Evaluation of optical flare and its effects on spectral estimation accuracy, in: Technical Report - Munsell Color Science Laboratory, Rochester.

Day, E., 2003b. The effects of multi-channel visible spectrum imaging on perceived spatial image quality and colour reproduction accuracy, Munsell Color Science Laboratory, Rochester Institute of Technology.

Dupuis, G., Elias, M., Simonot, L., 2002. Pigment identification by Fiber-Optics Diffuse Reflectance Spectroscopy. *Applied Spectroscopy*. 56 **10**, 1329–1336.

Eastman Kodak, 1970. Kodak filters for scientific and technical uses. Eastman Kodak: Rochester, New York.

Eastman Kodak, 1972. *Ultraviolet and Fluorescence Photography*, Eastman Kodak, Rochester, New York.

Frei, R., MacNeil, J., 1973. *Diffuse Reflectance Spectroscopy in Environmental Problem Solving*. CRC Press: Cleveland, Ohio.

Gargano, M., Ludwig, N., Milazzo, M., Poldi, G., Villa, G., 2005. A multispectral approach to IR reflectography, in: Art '05 – 8th International Conference on Non-Destructive Investigations and Microanalysis for the Diagnostics and Conservation of the Cultural and Environmental Heritage., Lecce, Italy, A148.

Geladi, P., Burger, J., Lestander, T., 2004. Hyperspectral imaging: calibration problems and solutions. *Chemometrics and Intelligent Laboratory Systems.* **72**, 209–217.

Greiner, R., 2000. The role of CCD cameras in astronomy, North Central Region of the Astronomical League.

Guntupalli, R., Grant, J., 2004. CCD Cameras tune in to scientific imaging, *Photonics Spectra.* **38**, 63–64.

Hamasaki, M., Ochi, S., 1996. Structure and operation of CCD image sensor. In: *Charge-Coupled Device Technology.* Overseas Publishers Association: Amsterdam.

Hapke, B., 1993. Combined theory of reflectance and emittance spectroscopy. In: *Remote Geochemical Analysis: Elemental and Mineralogical Composition.* Cambridge University Press: Cambridge.

Hardy, A., Perrin, F., 1932. *The Principles of Optics,* McGraw Hill: New York.

Harrick Scientific: What is Kubelka Munk? 2006.

He, X., Torrance, K., Sillion, F., Greenberg, D., 1991. A comprehensive physical model for light reflection, computer graphics. *Proc. of ACM SIGGRAPH.* **91**, 175–186.

Ichoku, C., Karnieli, A., 1996. A review of mixture modeling techniques for sub-pixel land cover estimation. *Remote Sensing Reviews.* **13**, 161–186.

Imai, F., Rosen, M., Berns, R., 2000. Comparison of spectrally narrow-band capture versus wide-band with a priori sample analysis for spectral reflectance estimation. In: IS&T/SID Eight Color Imaging Conference, Scottsdale, AZ, 234–241.

Imai, F., Taplin, L., Day, E., 2002. Technical report: Comparison of the accuracy of various transformations from multi-band images to reflectance spectra as part of end-to-end colour reproduction from scene to reproduction using spectral imaging.

Imai, F., Taplin, L., Day, E., 2003. Technical report: Comparative study of spectral reflectance estimation based on broad-band imaging systems. In: Munsell Color Science Laboratory, Rochester.

Johnston, R., 1973. Colour theory. In: *Pigment Handbook Vol. 3: Characterization and Physical Relationships.* John Wiley & Sons: New York.

Johnston-Feller, R., 2001. Color science in the examination of museum objects, Getty Conservation Institute, Los Angeles.

Kajiya, J., 1985. Anisotropic reflection models, computer graphics. *Proc. of ACM SIGGRAPH 85.* **19**(3), 15–21.

Kubelka, P., Munk, F., 1931. Ein Beitrag zur Optik der Farbanstriche, *Zeitschrift fuer technische Physik.* **12**, 593–601.

Kubik, M., Looking behind Kelly's Helmet: The materials and techniques of Sidney Nolan., AICCM Bulletin (In print).

Lee, A. E., Kubik, M., Creagh, D., Batterham, I., 2005. Kremer Pigments volume 1–3: Raman, FTIR and Fluorescence Spectra.

Leona, M., Winter, J., 2001. Fiber optics reflectance spectroscopy: A unique tool for the investigation of Japanese paintings, *Studies in Conservation.* **46**, 153–162.

Liew, S., 2001. Principles of Remote Sensing. Online tutorial, National University of Singapore.

Llyn, E., 1967. *Sidney Nolan: Myth and Imagery,* MacMillan: Melbourne.

Loebbert, G., 1992. Phthalocyanines. In: *Ullman's Encyclopedia of Industrial Chemistry,* VCH Verlagsgesellschaft: Weinheim.

Mairinger, F., 2000a. The Infrared examination of paintings. In: *Radiation in Art and Archeometry,* Elsevier: Amsterdam.

Mairinger, F., 2000b. The Ultraviolet and fluorescence study of paintings and manuscripts. In: *Radiation in Art and Archeometry,* Elsevier: Amsterdam.

Mansfield, J., Attas, M., Majzels, C., Cloutis, E., Collins, C., Mantsch, H., 2002. Near infrared spectroscopic reflectance imaging: a new tool in art conservation, *Vibrational Spectroscopy.* **28**, 59–66.

Martinez, K., Cupitt, J., Saunders, D., Pillay, R., 2002. Ten years of art imaging research. *Proceedings of the IEEE.* **90**(1), 28–41.

McLaren, K., 1983. *The Colour Science of Dyes and Pigments,* Hilger, Bristol.

Melessanaki, K., Papadakis, V., Balas, C., Anglos, D., 2001. Laser induced breakdown spectroscopy and hyperspectral imaging analysis of pigments on an illuminated manuscript. *Spectrochimica Acta Part B.* **56**, 2337–2346.

Melles Griot, 1999. *The Practical Application of Light,* Melles Griot: Irvine, California.

Moore, S., 1998. The Restoration of paintings by Sir Sidney Nolan, *Journal of the Association of British Picture Restorers* 13 Spring. 13–15.

Morgans, W., 1982. *Outlines of Paint Technology. Vol. 1: Materials*. Charles Griffin: London.

Nassau, K., 2001. *The Physics and Chemistry of Color*, Wiley Interscience, New York.

Newton: Optics; or, A treatise of the reflections, refractions, infectious & colours of light/based on the 4th ed., London 1730. Dover, New York. 1952.

Nolan, S., 1943a. [Box 6/18 John's letters to Sunday MS 13186) File 11a, in: John and Sunday Reed Archive - SLV, Melbourne.

Nolan, S., 1943b. [Box 6/18 file 11 (1 of 2)], in: John and Sunday Reed Archive - SLV, Melbourne.

Nolan, S., 1943c. [Box 2A/18 file 21 part 2 of 3 MS 13186], in: John and Sunday Reed archive, SLV, Melbourne.

Nolan, S., 1943d. [Box 6/18 John's letters to Sunday MS 13186) File 11c, in: John and Sunday Reed Archives - SLV, Melbourne.

Nolan, S., 1943e. [Box 7/18 file **15**, SN to JR], in: John and Sunday Reed Archive - SLV, Melbourne.

Nolan, S., 1943f. [Box 9/18 PA1168 File 8 SN to JR & others], in: John and Sunday Reed Archives - SLV, Melbourne.

Nolan, S., 1947. [PA 1168 Box 2/8 file 6], in: John and Sunday Reed Archive - SLV, Melbourne.

Nolan, S., 1948. [Box 2A/18 File 21/Part 1 of 3 PA 1168], in: John and Sunday Reed Archive - SLV, Melbourne.

Nolan, S., 1962. Conversations with Sidney Nolan/Interviewed by Hazel de Berg, [Sound recording].

Nolan, S., 1980. Interview with Sidney Nolan by Lois Hunter, [Sound recording].

Nolan, S., 1987. Interview with Sidney Nolan by A Durham, National Gallery of Australia, 30 January 1987, [transcripts].

Oltrogge, D., Hahn, O., Fuchs, R., 2002. Non-destructive analysis of the Codex Egberti. In: Art 2002: 7th International Conference on Non-destructive Testing and Microanalysis for the Diagnostics and Conservation of the Cultural and Environmental Heritage, Antwerp, Belgium.

Payne, G., Wallace, C., Reedy, B., Lennard, C., Schuler, R., Exline, D., Roux, C., 2005. Visible and near-infrared chemical imaging methods for the analysis of selected forensic samples. *Talanta*. **67**, 334–344.

Pelagotti A., Pezzati, L., Bevilacqua, N., Vascotto, V., Reillon, V., Daffara, C., A study of UV fluorescence emission of painting materials. In: Art '05 – 8th International Conference on Non-Destructive Investigations and Microanalysis for the Diagnostics and Conservation of the Cultural and Environmental Heritage. Lecce, Italy, 2005, A97.

Pellegri, P., Schettini, R., 2003. Acquisition and mosaicking of large/high-resolution multi-/hyper-spectral images. In: Tecnologie Multispettrali, Aspetti Teorici ed Applicativi, Quaderni di Ottica e Fotonica, Parma, 25–39.

Picollo, M., Porcinai, S., 1999. Non-destructive spectroscopic investigations of artworks, *Recent Res. Devel. Applied Spectroscopy* **2**, 125–135.

Prost, G., 2001. *Remote Sensing for Geologists: A Guide to Image Interpretation*, Taylor & Francis: New York.

Roberts, D., Gardner, M., Church, R., Ustin, S., Scheer, G., Green, R., 1998. Mapping Chaparral in the Santa Monica Mountains using multiple endmember spectral mixture models. *Remote Sensing of Environment*. **65**, 267–279.

Roselli, I., Testa, P., 2005. High Resolution VIS and NIR Reflectography by digital CCD telescope and imaging techniques: Application to the fresco 'Vergine con Bambino' in S. Pietro in Vincoli, Rome, in: Art '05 - 8th International Conference on Non-Destructive Investigations and Microanalysis for the Diagnostics and Conservation of the Cultural and Environmental Heritage., Lecce, Italy, A29.

Rosen, M., Fairchild, M., Ohta, N., 2002. An introduction to data-efficient spectral imaging, in: CGIV 2002: Proceedings of the first European conference on color in graphics, imaging and vision. Poitiers, 497–502.

Rosen, M., Jiang, X., Lippmann 2000: A Spectral image database under construction. In: International symposium on MI and color reproduction for digital archives., Chiba University, Japan, 1999, 117–122.

Saunders, D., Cupitt, J., 1993. Image processing at the National Gallery: The Vasari Project. *National Gallery Technical Bulletin*. **14**, 72–85.

SBIG: Users guide: CCDOps version 5, Santa Barbara Instrument Group, Santa Barbara, CA, 2003.

Scholten, J., Klein, M., Steemers, A., de Bruin, G., 2005. Hyperspectral imaging - A Novel non-destructive analytical tool in paper and writing durability research., in: Art '05 - 8th International Conference on Non-Destructive Investigations and Microanalysis for the Diagnostics and Conservation of the Cultural and Environmental Heritage., Lecce, Italy, A68.

Slawson, R., Ninkov, Z., Horch, E., 1999. Hyperspectral imaging: Wide-area spectrophotometry using a liquid-crystal tunable filter. *Astronomical Society of the Pacific*. **111**, 621–626.

Specifications for Artists' Oil and Acrylic Emulsion Paints. In: Annual Book of ASTM Standards, ASTM, Philadelphia, 1985.

Tate, Paint and painting: An exhibition and working studio sponsored by Winsor & Newton to celebrate their 150th anniversary, Winsor & Newton/Tate Gallery, London, 1982.

Torre, S., Rosina, E., Catalano, M., Faliva, C., Suardi, G., Sansonetti, A., Toniolo, L., 2005. Early detection and monitoring procedures by means of multispectral image analysis., in: Art '05 - 8th International Conference on Non-Destructive Investigations and Microanalysis for the Diagnostics and Conservation of the Cultural and Environmental Heritage., Lecce, Italy.

Turner, G., 1967. *Introduction to Paint Chemistry*, Chapman & Hall: London.

Vallee, O., Soares, M., 2004. *Airy Functions and Applications to Physics*. Imperial College Press: London.

van der Weerd, J., 2002. Microspectroscopic analysis of traditional oil paint, Natuurwetenschappen, *Wiskunde en Informatica*, University of Amsterdam.

van der Weerd, J., van Veen, M., Heeren, R., Boon, J., 2003. Identification of pigments in paint cross sections by reflection visible light imaging microscpectroscopy. *Analytical Chemistry*. 75 **4**, 716–722.

Wendland, W., Hecht, H., 1996. *Reflectance Spectroscopy*, Wiley Interscience: New York.

Wilcox, R., 1997. *Introduction to Robust Estimation and Hypothesis Testing*, Academic Press: San Diego, California.

Williams, K., Whitley, A., Dyer, C., 1995. Applications of Raman microscopy and Raman imaging. In: *Frontiers in Analytical Chemistry*, Royal Society of Chemistry: London.

Wu, C., Murray, A., 2003. Estimating impervious surface distribution by spectral mixture analysis. *Remote Sensing of Environment*. **84**, 493–505.

Zhao, Y., Taplin, L., Nezamabadi, M., Berns, R., 2004. Methods of spectral reflectance reconstruction for a Sinarback 54 digital camera. In: Technical report - Munsell Color Science Laboratory, Rochester Institute of Technology.

Author Index

Subject Index

271

Printed and bound by CPI Group (UK) Ltd, Croydon, CR0 4YY

08/05/2025

01864806-0008